W0037335

Synthesis Lectures on Engineering, Science, and Technology

The focus of this series is general topics, and applications about, and for, engineers and scientists on a wide array of applications, methods and advances. Most titles cover subjects such as professional development, education, and study skills, as well as basic introductory undergraduate material and other topics appropriate for a broader and less technical audience.

Sanjay Kumar

A New Theory of Additive Manufacturing

Solvable and Unsolvable Manufacturing Problems

 Springer

Sanjay Kumar
Swarup Niwas
Gumla, Jharkhand, India

ISSN 2690-0300   ISSN 2690-0327   (electronic)
Synthesis Lectures on Engineering, Science, and Technology
ISBN 978-3-031-75426-5   ISBN 978-3-031-75427-2   (eBook)
https://doi.org/10.1007/978-3-031-75427-2

© The Editor(s) (if applicable) and The Author(s), under exclusive license to Springer
Nature Switzerland AG 2025

This work is subject to copyright. All rights are solely and exclusively licensed by the Publisher, whether the whole
or part of the material is concerned, specifically the rights of translation, reprinting, reuse of illustrations, recitation,
broadcasting, reproduction on microfilms or in any other physical way, and transmission or information storage
and retrieval, electronic adaptation, computer software, or by similar or dissimilar methodology now known or
hereafter developed.
The use of general descriptive names, registered names, trademarks, service marks, etc. in this publication does
not imply, even in the absence of a specific statement, that such names are exempt from the relevant protective
laws and regulations and therefore free for general use.
The publisher, the authors and the editors are safe to assume that the advice and information in this book are
believed to be true and accurate at the date of publication. Neither the publisher nor the authors or the editors give
a warranty, expressed or implied, with respect to the material contained herein or for any errors or omissions that
may have been made. The publisher remains neutral with regard to jurisdictional claims in published maps and
institutional affiliations.

This Springer imprint is published by the registered company Springer Nature Switzerland AG
The registered company address is: Gewerbestrasse 11, 6330 Cham, Switzerland

If disposing of this product, please recycle the paper.

Contents

About the Author

Dr. Sanjay Kumar is the author of four books: *Additive Manufacturing Processes*, 2020 (https://www.springer.com/gp/book/9783030450885), *Additive Manufacturing Solutions* (https://www.springer.com/gp/book/9783030807825), 2021, *Additive Manufacturing Classification*, 2022 (https://www.springer.com/book/9783031142192), and *Additive Manufacturing Advantage*, 2023 (https://link.springer.com/book/9783031345623).

Introduction

1

1.1   What Is Additive Manufacturing (AM)?

In AM [1–4], a part is visualized as a set of horizontal layers. Therefore, AM requires either to have layers and a method to join them or a method to make and join them. To have layers means to bring them from somewhere if that is required. In almost all AM techniques, making a layer is accompanied by its automatic joining with an underlying layer or a platform, implying that making a layer means both making and joining.

AM simplifies fabrication by reducing it to making layers and joining them. It is called simple on the assumption that making layers and joining them are more simple than making the whole part at once. If this is not the case, it is not simple. For example, if a cube is made from a bigger cube by machining, it is simple. There is no more scope left to again simplify it. If AM is used to make this cube layer by layer, AM does not simplify it. On the contrary, AM makes it complex—a number of layers need to be arranged and joined, while machining requires only trimming a bigger cube. Thus, only when a part is complex can AM be invoked to simplify the complexity.

AM relies on a method to simplify the fabrication. The method is to fabricate layer by layer. The method relies on the fact that the actual happening of making layers and joining them is a simplification. It means a part being complex is an essential requirement to invoke AM, but it is not a sufficient requirement. The sufficient requirement is that making layers and joining them must not turn out to be more complex than the complexity of a part.

© The Author(s), under exclusive license to Springer Nature Switzerland AG 2025
S. Kumar, *A New Theory of Additive Manufacturing*, Synthesis Lectures on Engineering, Science, and Technology, https://doi.org/10.1007/978-3-031-75427-2_1

1.2   What Are Advantages?

All advantages of AM are derived from its fundamental characteristic, i.e., freedom to add material layer by layer (Fig. 1.1). Though the freedom is restricted to add in a plane before moving to add in the next plane, the freedom left after the restriction is enough to lead to advantages.

The advantages are

1. Freedom to design—ability to make products of various designs. It gives three sub-advantages: freedom to save material, freedom to save energy, and freedom to make multi-material parts,
2. Freedom to move—ability to be used at various (including non-conventional) places due to its versatile and compact systems, and

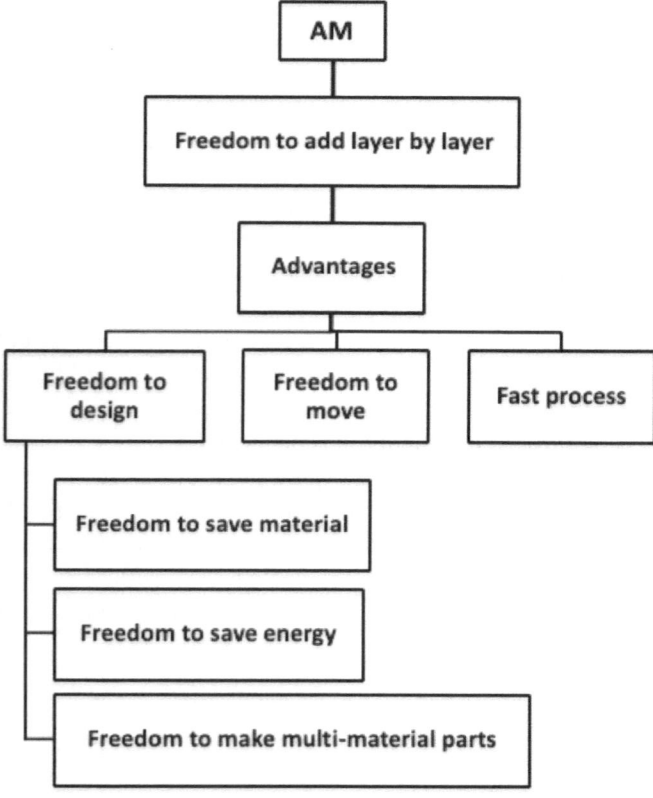

Fig. 1.1 Advantages

3. Fast process—ability to manufacture fast, as the manufacturing is direct and not through design-specific tools.

1.3   What Is Freedom in 'Freedom to Design'?

Conventional manufacturing (CM) relies on tools to make parts. If many types of parts are made, it is because there are various types of tools available to make them. And if many types of parts are not made, it is because these are the tools that become obstacles when they are used. Therefore, if there is freedom from the tools, the parts that could not be made due to the tools can be made. AM does not rely on those tools on which CM relies; therefore, AM has the freedom to make or design those parts without being restricted by those tools.

The freedom does not exist when AM is compared with non-conventional manufacturing (NCM) because NCM does not use traditional tools as well. Examples of NCM are electrical discharge machining, electrochemical machining, laser beam welding, plasma beam welding, laser beam machining, etc. Out of these processes, some processes such as laser welding, electron beam welding, etc. are additive processes. When these NCM additive processes manufacture layer by layer, they become AM. Therefore, AM does not have more freedom than NCM additive processes. On the contrary, they can have more freedom as they are not constrained to work layer by layer.

1.4   Why Does a Machining Tool Not Allow to Make Complex Parts?

There are several reasons; some of them are as follows:

- The tool cannot be made thin enough to make thin features. In extreme cases, if it is made, it can break while making the feature; or if it is not made thin, it may break the feature.
- A sharp corner depends upon the sharpness of a tool. If a sharp tool cannot be made and operated, there is no tool available to make a sharp corner.
- Making a complex part requires a number of tools that need to be arranged in a sequence of operations, i.e., a decision needs to be made about which tool will come after the use of which tool. This does not allow to make all types of parts.

  For example, a small milling tool can make a fine feature, but it cannot make the same at the inside base of a hollow cylinder because it cannot access the base. Hence, the sequence of operation, i.e., to make a hollow cylinder using a big milling tool and then using the small milling tool to carve a feature will not work. The sequence of operation in the reverse order will not work either because a cylinder is made by removing the material (i.e., clearing the feature). Therefore, the complex part cannot be made not because each tool does not have ability to make a shape. But because when the operations of the two tools are combined, the result of the operation of one tool does not allow another tool to work.

- Making a complex part may require excessive material removal. If the part is formed from a difficult-to-cut material, the tool wear will further increase. In the absence of a suitable tool for such materials, the part will not be formed.
- The geometry of a part is not free from the type of a machining tool. For example, a drilling tool cannot make a feature free from a cylindrical shape. If a non-cylindrical shape needs to be made with a drilling tool, it is made with difficulty.

1.5   If AM Makes Simple Products, Whether It Is an Advantage?

When AM makes a simple product, there are two things involved. First, a process to fabricate, and second, a product that is an outcome. The outcome is not an advantage because other techniques can also provide it. But since the outcome is due to AM, this can lead to other advantages.

1.6   What Are the Advantages When Simple Products Are Made?

Making simple products by AM will be advantageous if making them are economic, fast, or if they are made at such places where other techniques are not suitable to be used. These advantages can be summarized as economic, fast, and places to fabricate at. The first advantage, i.e., economic, is temporary. Therefore, it is not included. The second advantage is included as fast process. The third advantage is included as freedom to move.

Why is the first advantage temporary? If AM makes a simple product economically, it is due to two reasons: (1) the condition of the market when the product is made and when it is out for sale, and (2) the technical aspect of a manufacturing technique. Therefore, the second reason, i.e., the technique, is not the only reason for a product to be economic. The technique cannot claim that it solely provides the advantage. Even if it claims, the claim is time-dependent because the market condition will not always be the same. Therefore, even if AM makes a simple product economically, this advantage has no guarantee to last long.

The second advantage is already recognized when AM is called rapid prototyping or rapid manufacturing. If the product is simple, the product is not benefitted from the design advantage of AM. Nonetheless, if the product is required to be made fast, its fabrication is an advantage.

1.7   What Is Freedom to Move?

What AM can manufacture does not completely show what it is. What an AM system is can help show what AM is. If AM can be used on an office desk, it has ability to be packaged in such a way that it can work in an office environment. That is, it must have some qualities as follows: it does not have chemicals that create fumes; or it does not consist of vibrating components that create disturbances; or its size along with its subsidiary equipment is not too big to be kept only in a workshop; or it does not require special facilities that can be found only in a workshop, such as pressurized gas; or it does not have delicate components that demand extra precaution, asking for an environment different than that of an office. Though, not all AM techniques have such qualities. For example, an AM system equipped with a high power laser cannot be kept in an office environment, but an AM system that uses plastic filaments can be used. Similarly, an AM system, i.e., powder bed fusion, can be used in a remote area to make high-value metallic parts, while an AM system using filaments is not suitable for that purpose.

Thus, an AM system that is used in an office environment is not to make all products that AM can make, but the system is still useful because the types of designs it can convert is not possible by conventional systems. This makes AM versatile, which allows it to be utilized at different places. This is a type of freedom—the freedom to move or to be located.

1.8   What Is the Difference Between a Simple and a Complex Product?

A simple product has a simple design, e.g., a hollow cylinder; while a complex product has a complex design, e.g., a turbine blade.

A hollow cylinder has a simple design; therefore, it can be fabricated easily. Vice versa, if the cylinder can be fabricated easily, it is a confirmation that its design is simple for AM. To verify its fabrication, it can be checked how an extrusion deposition system will fabricate it. The system's nozzle will deposit a material by moving in a circle and then repeating the step by moving up—this is simple requiring simple, predictable data. The data will consist of the diameter of the circle and the height of each increment— both values remain same for each layer. Therefore, there is no need to make a design, the operator can visualize it without the help of software. Again, there is no need for software to convert the design into machine-readable instructions; the operator can provide the instructions by inputting the data.

What if the operator is asked to make a turbine blade? The operator cannot visualize the design in details because the design has a number of channels and curves that change in dimension with each layer. These will create enormous data that are difficult to operate. Thus, when design becomes complex, the difficulty of the operator increases. This is the difference between a simple and a complex product. The difficulty faced by an operator is a measure for the complexity of the product. If the operator faces more difficulty, the product is more complex.

The difficulty can also increase when the material is not compatible giving inadequate bonding, or the nozzle does not move accurately giving inaccurate dimension. But when the difficulties due to machine or material are overcome, there is still difficulty if the turbine blade instead of a cylinder is made. This shows that the difficulty of an operator can be related to an increase in the design complexity. This difficulty is not related to the skill or experience of an operator but is equal to the amount of tasks that are needed to be performed. If the design is simple, tasks are few, i.e., machine-readable instructions are few. When the design becomes complex, instructions are more.

What if the same cylinder, which is an upright orientation, is made in a horizontal orientation (Fig. 1.2). Then, instructions will be more because the dimension of the layers will change with a change in height. More instructions will not lead to a complex product but the same product. This demonstrates that complexity in fabrication does not lead to complexity in design. This demonstration seems to contradict that the complexity of a product must be related to the number of instructions. But this demonstration clarifies that the number of instructions changes with a change in process steps, material, orientation, etc., therefore, equating the number of instructions to complexity in design will be misleading unless the fabrication is optimized. When the fabrication for a cylinder is optimized, the cylinder only in an upright orientation will be selected to compare with the turbine blade.

Considering a situation where making an upright cylinder is more difficult than making a turbine blade in an AM technique. Then which one is simple? Since it is the difficulty in fabrication that decides which design is simple or complex, the turbine blade will be a simple product while the cylinder will be a complex product. Thus depending on the

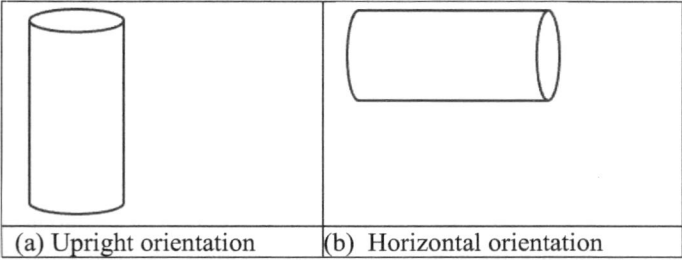

| (a) Upright orientation | (b) Horizontal orientation |

Fig. 1.2 Cylinder orientations

type of AM techniques, a design or product can be simple in one AM technique while complex in another AM technique. But the reason why it is simple or complex will be the same in all AM techniques.

1.9   What Is Complexity?

Complexity is equal to fabrication problems. If there is no problem for fabricating a complex part, the part is not complex. A geometry is simple or complex depends upon whether fabricating it is easy or difficult. If a part is complex in other techniques to make, it does not imply that the part is also complex in AM when there is no problem in fabricating it in AM. Similarly, if a part is easy to make it in AM, it does not imply that the part will not be complex in other techniques.

1.10   Why Is AM Itself Not an Advantage?

If there is no difficulty in converting a design into a product (e.g., a hollow cylinder), there is no difficulty available that can be removed by using AM. Then AM cannot demonstrate its advantage of how it simplifies a complex design into a set of simple steps. But AM can still be used if it is better than conventional techniques—this is the advantage of AM over other techniques. This advantage only shows how the conversion of a design is simplified in AM. The simplification has limited value if the conversion is limited only to simple designs. This simplification has no value if the conversion is limited only to simple designs while the conversion is sought for complex designs. Therefore, the simplification, which is freedom to add materials layer by layer, is more an attribute than an advantage.

If this freedom leads to make complex products, this is an advantage. If this freedom leads to make simple products, this is also an advantage; but this leads to disadvantages as well. For example, if AM makes a big metallic square block that is made economically and fast by machining, then using AM is a disadvantage. Therefore, this freedom, which is AM, is itself neither an advantage nor a disadvantage. That is why it is not shown as an advantage in Fig. 1.1.

1.11   What Are the Two Ways in Which Material Is Saved?

AM saves material in two ways: (1) by default, and (2) by design.

- **By default**: Material is added where it is needed in AM. It means more material than what is added is not required. In many techniques, more materials are required. For example, in machining, a block bigger than a part is always required. This 'more material' is saved when AM is used. This is material saving by default. In machining, 'more material' will be saved only when the design of a part is similar to the design of a block from which the part is intended to be formed. That is, material is saved in machining only when a finished product that does not require machining is brought to be machined. Therefore, the machining process does not allow to save material and to machine at the same time. That is, in principle, it is not possible to save material in machining. While it is possible to save material in AM, in principle. If it is said some machining saves material, it implies that some machining wastes less material.

Thus saving by default is not an advantage due to what AM does but due to what AM is. There is no effort done by AM to have this advantage. Material saving by default is not an outcome over which AM has control. AM does not have option to change this outcome, i.e., it cannot increase or decrease the amount of the material that is saved using this way. Thus, there is no freedom to save material. It is because freedom occurs when it is left up to users to exercise it. For example, if material is intended to be saved by changing the design of a part, then it depends on users to change the design and save material. This freedom does not exist when the design is already selected and converted. That is, when the design is converted, the material that is saved cannot be changed.

What if this saving is considered a disadvantage; what if AM is expected not to save this material because there is no facility to store this material, even then AM cannot work differently to consume more material. Because AM works only in one way, i.e. by adding the material where it is needed. If the way AM works is not considered an advantage, it will not be an advantage but will be a perspective. It is considered an advantage because it is observed from the perspective of other techniques. For example, it is an advantage only because what the amount of material machining consumes is considered normal. If it were considered waste rather than normal, the amount of material AM consumes would have been considered normal; then there would not have been any material saving by default. Therefore, if material is saved by default in AM, it is not the advantage that is gained by the application of AM as a manufacturing tool. Therefore, material saving by default is not included as an advantage in Fig. 1.1.

- **By design**: Material can be saved if a part consists of less material than what it previously required. For this, the design of the part needs to be improved so it will consist of less material when converted into a part. For conversion, there requires a technique that is able to do it. If AM does it, AM has the ability to save material more than it saves by default. This ability is freedom to save material.

  What if machining also has the ability to convert this improved design? Then machining will also help make a part using this improved design. But it does not imply that machining will also be able to save material. Because machining does not work on a principle that says making a thinner section means less material addition. On the contrary, machining will remove more material if a thinner section is required. It is possible that an improved design leads to a design that has a smaller dimension, which leads to a smaller workpiece to start with. This will lead to material saving because smaller workpiece means less material loss after machining. But this material saving is due to the requirement of a smaller workpiece and not due to the requirement of less material to machine a thinner section.

If AM is not able to convert an improved design, AM has no ability to save material by exercising its quality to convert an improved design. If machining is not able to convert the improved design as well, machining also has no ability to save material by exercising its quality to convert the improved design. Then AM is same as machining as both are not able to convert the improved design. But AM can still save material by using an old design instead of an improved design. This material saving, i.e., by default, shows that there can still material be saved even if AM fails to be better than machining. This material saving, thus happened, shows not the advantage but the disadvantage of AM. This advantage of AM (i.e., this material saving) is the result of the main disadvantage of AM (i.e., inability to convert an improved design).

When AM has ability to convert an improved design, i.e., when AM has ability to make a part that is called complex when compared with other techniques, this ability is freedom to design. When AM has ability to convert an improved design, this can or cannot lead to material saving. When this leads to material saving, this is freedom to save material. Thus, the freedom to save material is not an independent freedom but depends on the freedom to design. It means the freedom to save material follows the freedom to design, which is shown in Fig. 1.1 where the former freedom is the sub-advantage of the latter freedom. Thus, AM cannot save material without making a complex part unless the material it saves is saved by default.

1.12 Does Freedom to Save Material Lead to Actual Material Saving?

Freedom to save material means AM allows to save material by converting a design into a part that consists of less material. During fabrication, material may spill or deteriorate. This will change the expected material saving. Thus, a part that consists of less material may not have required less material to fabricate.

1.13 About the Book

AM is an arrangement of layers. Changing the position of layers during layer deposition changes the arrangement. When the next layer is shifted with respect to the last layer, the shift can create a gap. When consecutive gaps are created, this gives rise to an unsolvable manufacturing problem. This problem can become a basis to divide layer arrangements into two types. The first type, which gives unsolvable manufacturing problems, is where these gaps are created. The second type, which does not give unsolvable manufacturing problems, is where these gaps are not created. Layer arrangements can be related to manufacturing problems, which is a new theory.

The book mentions two consecutive gaps, two types of manufacturing problems, two types of layer arrangements, two types of overhangs, two types of shifts, two orientations of stair-stepping, two process types, and the requirement of two layers. It is no wonder AM prefers number 2.

References

1. Kumar S (2020) Additive manufacturing processes. Springer, Cham
2. Kumar S (2021) Additive manufacturing solutions. Springer, Cham
3. Kumar S (2022) Additive manufacturing classification. Synthesis lectures on engineering, science, and technology. Springer, Cham
4. Kumar S (2023) Additive manufacturing advantage. synthesis lectures on engineering, science, and technology. Springer, Cham

A Layer Arrangement

2

2.1 What Is the Difference Between Layer Deposition and a Deposition Process or an AM Process?

Layer deposition is an action that informs about the deposition of a layer. This layer is part of the design of a part. Additive manufacturing (AM) makes a part when layer deposition takes place for all the layers of the design of the part.

Deposition process is a family of AM processes or techniques; processes and techniques are used interchangeably. The meaning of the process in a deposition process is not strictly a set of steps. Otherwise, any part that is formed in an AM process with a slight variation in forming will not belong to the same process.

Deposition process is a family of AM techniques that restricts how layer deposition can take place. It can take place only when there is a free space to be filled. It cannot take place when the material it deposits finds the space occupied.

When the process informs that the space should not be occupied, it presupposes the existence of another family of techniques that facilitates the layer deposition without providing a free space. This is what the bed process is, where the layer deposition does not require material to travel from one point to another to fill up the required space. Because the material exists everywhere on a plane. But, this is a problem. Because if the material exists everywhere, this is not what a layer should be. A layer means the material should exist only where the design suggests it to exist. This requirement demands that the unnecessary material be removed from the plane, so the resulting layer will be part of the layer deposition. If it is not removed immediately, promise should be given that this material should not be part of a final product. The bed process is not about immediacy. It is about the promise. It gives promise by marking the unnecessary material; the material that is marked is removed. This is the marking that allows the layer deposition to take place in the bed process. How the marking will take place changes from one technique

© The Author(s), under exclusive license to Springer Nature Switzerland AG 2025
S. Kumar, *A New Theory of Additive Manufacturing*, Synthesis Lectures on Engineering, Science, and Technology, https://doi.org/10.1007/978-3-031-75427-2_2

to another. Whether the marking will be accompanied by material consolidation changes again from one technique to another, but the bed process is not free from marking.

When unnecessary material is removed, this reminds of machining. For making a part, as the machining relies on removing unnecessary material, the bed process is also not free from the reliance on removing unnecessary material. Although, the degree of removal force required in machining and bed process do not match. Thus, the bed process is an AM equivalent of machining. In deposition process, material is removed as well, but it is not because material removal is required as part of the process. But because when the material could not be deposited well while executing the process, the inaccurate deposition needs to be corrected. Therefore, out of the deposition and the bed process, it is the bed process that is closer to the machining.

2.2   What Is a Layer in Bed Process?

In bed process, which layer is a layer? If it is said that a layer is a layer in a bed process, it is a correct statement. If it is said that a layer is not a layer in a bed process, it is again a correct statement. Which statement is correct when?

When it is said AM is a layer by layer technique, it means a design needs to be first divided into a number of layers. The existence of the layer notifies the existence of a design from which the layer originates. The layer is part of the design. The layer itself has a design. This layer is not that layer that is used to cover or protect a surface and does not have a design but must conform to the surface design. Layer deposition means deposition of only those layers that have a design, so they collectively make a product as per the design. When it is said that a layer is a layer in bed process, it means bed process is not free from the requirement of AM that says a layer is a layer because it has a design.

How does bed process take place? Material is deposited on a platform to make a bed. What is a bed? A bed is a layer of material deposited on a platform. When the layer of material is deposited, is the layer deposition taking place? If it is said that a layer is not a layer in the bed process, it reminds that a layer that is the synonym of a bed is not the layer that is part of a design. It can better be said that a layer that is a bed is the precursor of the layer of the design of a part. Thus, bed deposition is mentioned as layer deposition as well.

2.3   What Is the Difference Between Layer Deposition and Layer Arrangement?

Layer deposition is the deposition of a single layer or of many layers. Layer arrangement is an arrangement of layers. A design is an arrangement of layers. A part is an arrangement of layers. AM is an arrangement of layers.

When it is said a design is an arrangement of layers, it means a design does not work in AM unless it is divided into many layers. These layers are arranged in the design as per the design. If a design is different, the arrangement of layers in the design is different. This is how different parts are made. However, if the orientation of a design changes, it changes how layers are arranged. Therefore, for a design, there will be a number of arrangements as per the orientation (Fig. 2.1). The orientation of a design is more important than the design itself, because this is the orientation that informs which arrangements can allow to convert a design into a part conveniently. None of the arrangements (Fig. 2.1a–d) accurately fits the design, showing the limitation of manufacturing a design through layers.

When it is said a part is an arrangement of layers, it means a part is made from a design that is made up of many layers.

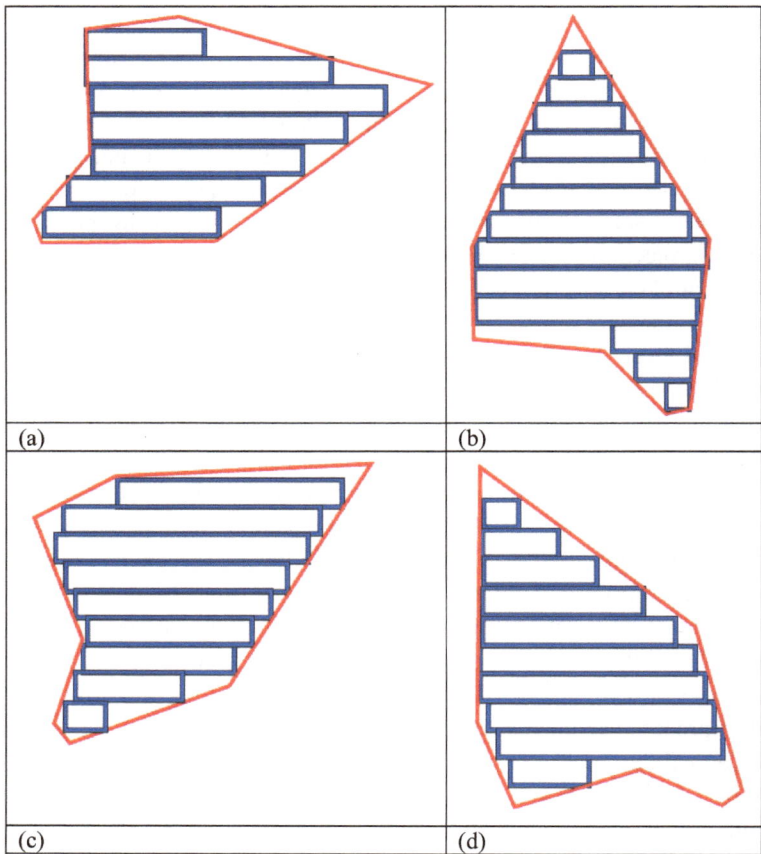

Fig. 2.1 Different layer arrangements for same design

When it is said AM is an arrangement of layers, it informs what AM does when it is used—it deposits one layer after another, giving rise to an arrangement of layers. But, AM has a restriction that it does not decide about the arrangement. The decision is taken when a design is made. AM is to make the replica of the arrangement that is in the design. This allows AM to be understood from the design point of view but not from the layer arrangement point of view. But, it is the layer arrangement point of view that is more important because there is no dearth of designs to be converted, but there is a dearth of manufacturing techniques that can make layer arrangements.

2.4   What Is a Layer? How Should It Be Defined?

A layer is the shortest unit of a design that is built in AM. It is usually defined by its thickness, shape, and area. The definition needs to be revisited.

When a design is divided into a number of layers, a layer is not meant to be different from its projection on a horizontal plane. If the design is simple, e.g., a rectangular block, the layer of any thickness gives the same projection, which does not create the difference between a layer of any thickness and its projection. If the design is not such simple, e.g., a sphere, the layer of any thickness does not give the same projection. For such a design, if a thick layer is selected, it gives problems because it has many projections. If the differences in projections are ignored and the layer is selected to manufacture a sphere, the result will not be a sphere. Therefore, a layer of any thickness can be selected as long as the selection does not create more than one projection. If the design is a sphere, the thickness of the layer should be selected to a minimum. So even if the selection creates more projections, their differences can be ignored. This will allow to make an approximate sphere. That is a sphere manufactured in AM. Can AM ignore the differences in order to manufacture? This is what AM is. It ignores because it has limitations. Its limitation is that it manufactures a layer through its single projection. It selects a single projection and consolidates through it. If a layer has two projections, it does not select the upper projection and consolidates through it to meet the lower projection. Because it has no ability to give a direction to the consolidation.

In some cases, consolidation gets a direction when the directional solidification takes place. But it has only one direction to take, and that is the direction that happens by default in AM. If it takes another direction than the default direction, it is due to the lack of control of process parameters and not because AM has ability to change the direction as per the vertical design of a layer.

Therefore, a layer is required to be defined by a single projection. It does not limit the thickness of a layer it can manufacture. Thus, a rectangular block can have a layer equal to the rectangular block, but a sphere cannot have a layer equal to the sphere. The block can have layers of various thicknesses without affecting its size and shape (Fig. 2.2).

Fig. 2.2 No change in
rectangular block

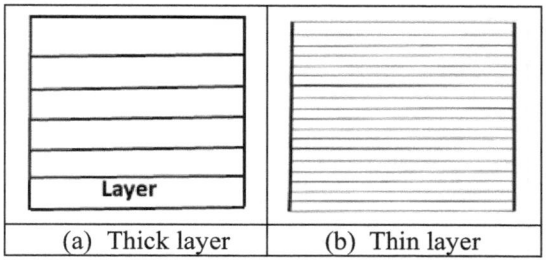

| (a) Thick layer | (b) Thin layer |

If a layer is a projection, which 3D shapes can a layer have? It can take only those 3D shapes that are the extension of the projection. A layer can be a circular disk—an extension of a circular area, a ring—an extension of a circular line, a rectangular block— an extension of a rectangular area, a square block—an extension of a square area, a star-shaped disk—an extension of a star-shaped area, etc., but it cannot be a pyramid, a trapezium, or a rhombohedron, etc.

Figure 2.3 shows the projections of various 3D layers on horizontal planes. When the shape is either rectangle or a circular disk, the projection is only one area (Fig. 2.3a, b). This is because the upper and the lower surface of the layer has the same area. For a trapezium layer, its upper and lower surfaces have different areas, which gives two projections, as shown by a dotted and a solid line (Fig. 2.3c). AM requires only one projection, assuming the consolidation will extend normal to the projection (Fig. 2.4). The downward consolidation gives the depth of the consolidation, which is a layer thickness. For a trapezium layer (Fig. 2.4c), when there are two projections, it is expected that AM will consolidate through the upper projection and reach the lower projection, i.e., there is a desire for the consolidation to have a direction, i.e., the extension of the consolidation should not be according to what AM does but what the design of the layer dictates. But the extension will be normal to the projection, making a rectangular layer rather than a trapezium layer.

Therefore, a layer is what its upper projection is, i.e., a single projection defines a layer in AM. What is the use of a trapezium layer when it acts like a rectangular layer; what is the use of its two projections when one of the projections is not recognized by AM? A 3D shape of a layer does not exist in AM unless is made by the extension of the 2D shape of a layer. When a trapezium layer is selected, the result will be a rectangular layer, which will create a gap that is the difference between the solid area of the rectangular area and that of the trapezium (Fig. 2.4c).

A gap is generally created when a layer selected does not match the curve of the design of a part; this gap is created because AM does not meet the expectation of the design— this is a fact. When a layer is selected that has more than one projection and the gap is created, it is not because AM does not meet the expectation but because the 3D design of the layer does not meet the expectation of AM—this is a fact. Both facts say the same

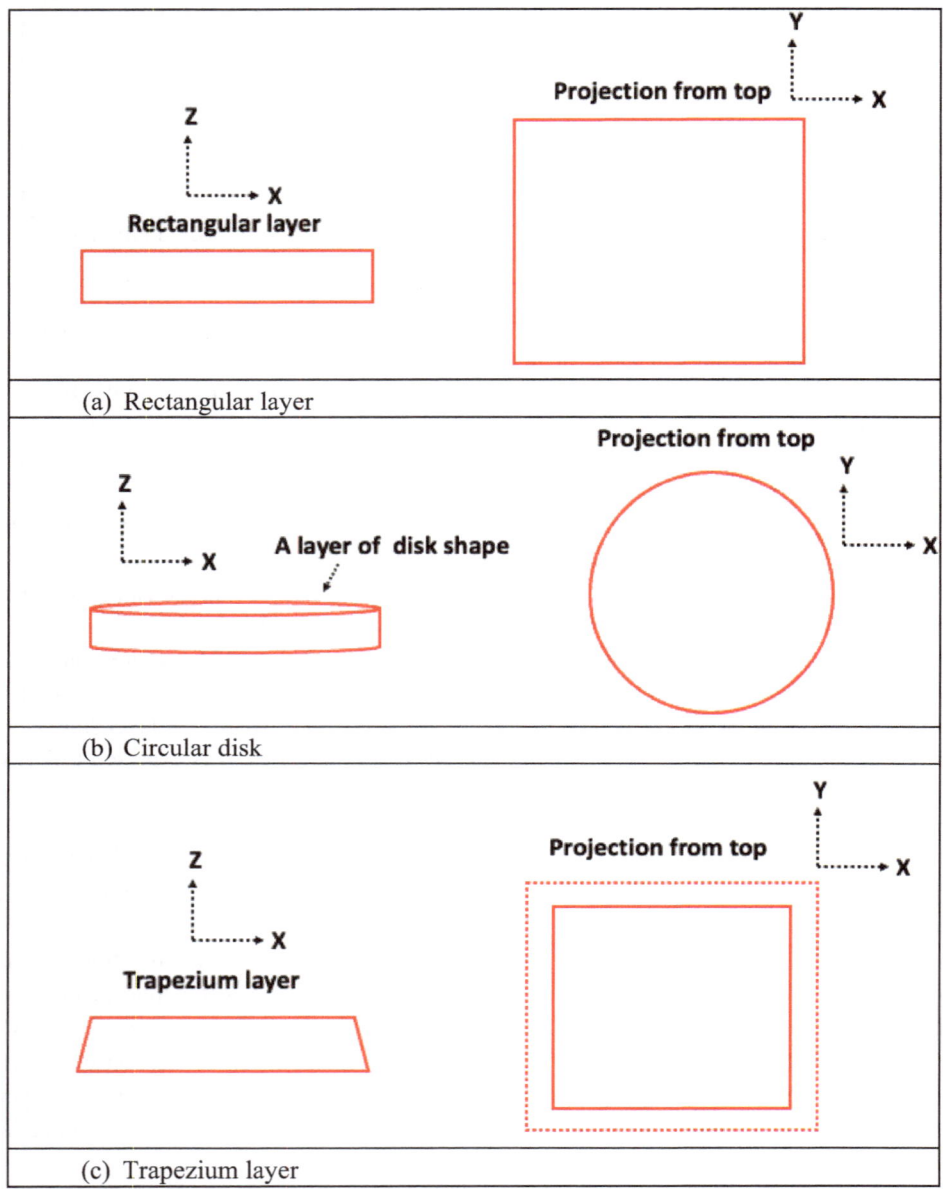

Fig. 2.3 Projection of layer of various 3D shapes

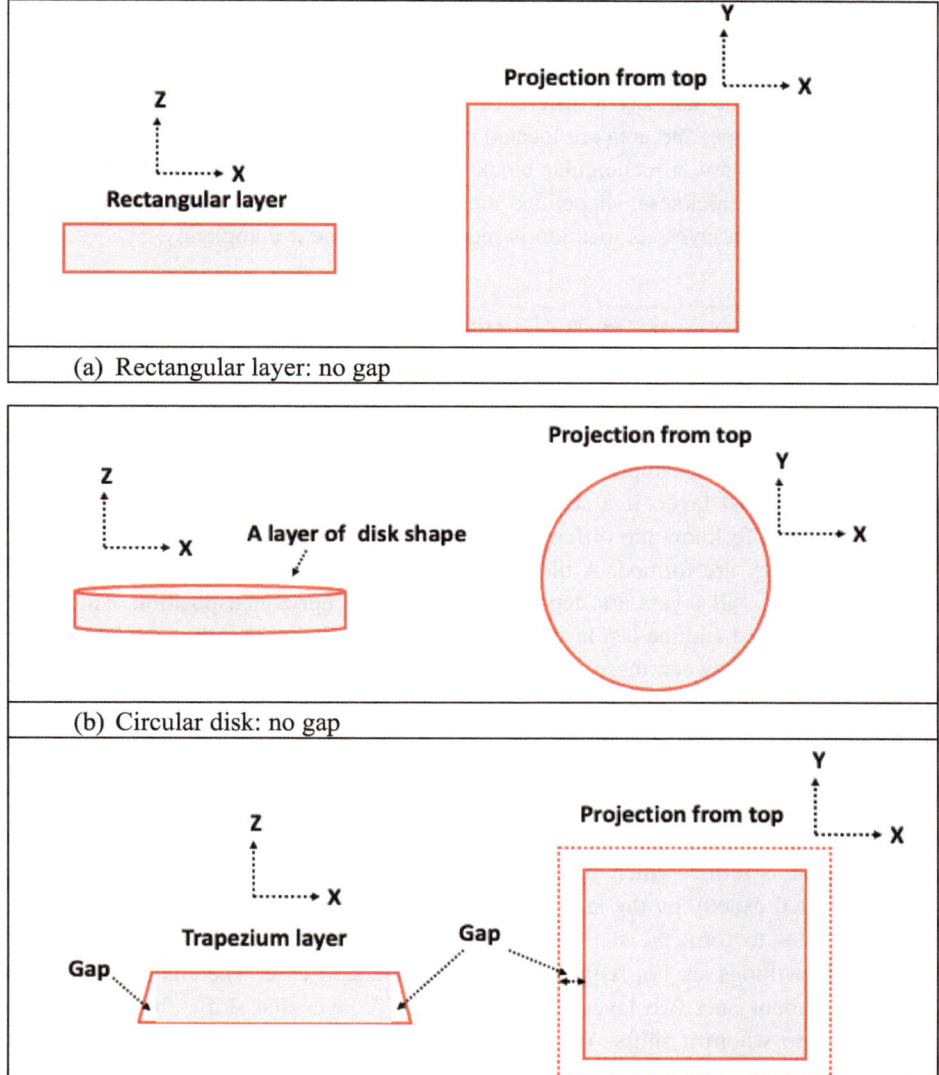

Fig. 2.4 Layer consolidation of various 3D shape

fact: if a layer does not match the design, there is no solution for it as the layer cannot be stretched to match.

The thickness, shape and the area of a layer is not sufficient to define a layer, as this information does not help say a difference between two designs where the layer of the same thickness, shape, and area are located differently. For example, this information will not be able to say how a rectangular block is different from a stair-case made from the layers of the same thickness, shape, and area (Fig. 2.5a, b). Thus, besides the thickness, shape, and area of a layer, its location is required to define it completely.

2.5   Can a Layer Be Defined Without a Coordinate?

A layer needs a coordinate to say where it is in a 3D space. But having coordinates for all layers means more information than required, because all layers are not needed to describe a design. For example, a rectangular block can be described by two layers, i.e., the first and the last layer. If a stair-case is made, positions of only two layers do not help (Fig. 2.5b). To know the difference between a block and a stair-case, it is required to know how they are formed. A block is formed when a layer is deposited exactly on the last layer, i.e., all layers are deposited on the same horizontal position. Thus, if the locations of the first and the last layer are known, it will inform that the locations of other layers are exactly between these two layers. Instead of two layers, if only the first layer is known, it will inform that other layers are exactly above this layer. Thus, only one layer can also describe how a rectangular block forms. It is because further layers that are deposited on this layer do not change their horizontal positions. But if they do not change, no shape other than the block can form. Due to the lack of change, the coordinates of many layers are not required, i.e., a coordinate is required only when there is a change.

A stair-case is formed when at some height (e.g., at height 1 and 2, Fig. 2.5) the layer is not deposited exactly on the last layer but at a shifted position. This is a change. This change initiates to form the step of a stair-case. If only the changes can be taken into account, other things are not required to describe the stair-case. The change requires the information about only two layers that it involves. A layer that shifts and another layer with respect to whom it shifts. What happens can be described by informing about the height at which it shifts. If one layer is a next layer, another is a last layer—using these two is not the replacement of coordinates, but these can help.

The transformation from a rectangular block to a stair-case can be described by what changes happens twice. And the changes can be described by how the next layer changes with respect to the last layer. In this case, the next layer shifts rightwards. If the next layer does not shift rightwards but becomes smaller in area, the size of the block will start to change (Fig. 2.5c). This change will make a sub-block if there is no change until further change. The difference between the two changes gives the height of the sub-block. The

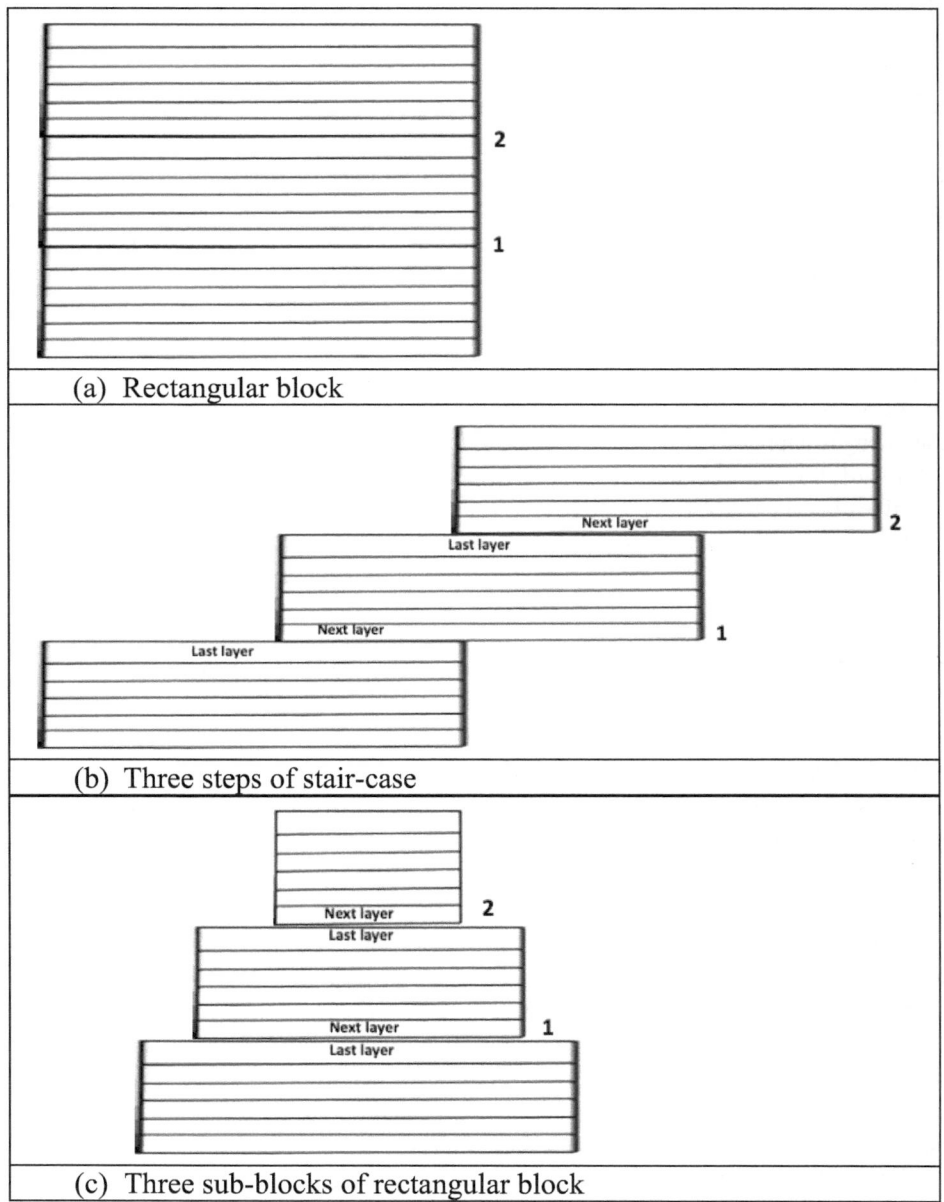

(a) Rectangular block

(b) Three steps of stair-case

(c) Three sub-blocks of rectangular block

Fig. 2.5 Difference between parts because of changes in next layer

change of the next layer at different heights helps to know how steps or sub-blocks form. Thus, without using coordinates, different designs can be explained.

Why should not layers have numbers, such as L1, L2, L3, etc. so a part geometry can be described by how L3 changes with respect to L2? Because layer deposition moves in a forward direction, which provides a benefit to know only one layer at a given height—this is the only coordinate required. If there are numbers for layers, at least serial numbers of two layers are required, i.e. L2 and L3. Besides, a direction is required to count numbers, i.e., from top or bottom. It means for L2, it needs to be ascertained which layer is going to be deposited, it is L1 or L3? If a layer is called the next layer, it informs about the forward direction of deposition. Figure 2.5b, c can be described by saying how the next layer changes with respect to the last layer at heights 1 and 2. When it changes at height 1, it means the changes at height 2 depends on the change that happened at height 1, because the deposition moves in a forward direction. When it changes at height 2, it means what the changes happened at height 1 cannot be unchanged, because the deposition moves in a forward direction, i.e. only deposition is possible; deposition and undeposition are not possible. It means layer deposition is not a shuffling of layers where any layer can be shuffled to give any layer arrangement. A layer is allowed to shuffle only when the layer is the next layer and only when the next layer shuffles with respect to the last layer.

2.6   How Will Defining a Layer Work in Different AM Processes?

When it is said that layer deposition goes forward, it is not only about the lack of space due to which the deposition cannot go backward, it is also about the constraint that does not allow it to go backward. In bed process, the deposition in the backward direction does not work because the process does not allow, while in deposition process, it can work if geometry allows.

In bed process, when the next layer shifts outside the boundary of the last layer, there is no connection between the next and the last layer. Then the part does not grow, i.e., part-1 stops forming and part-2 starts forming (Fig. 2.6). When the next layer shifts leftwards at the end of the formation of part-2, there is again no connection between part-2 and the next layer. The next layer has opportunity to be deposited on part-1 to make the part bigger. But the deposition cannot take place to make part-1 bigger, because the layer deposition moves forward and cannot move backward.

What is the meaning of when it is said the layer deposition moves forward? It moves forward means when a layer is deposited, this deposition is for forever. This deposition cannot be removed to make a space for further layer deposition to take place even if the deposition happens by mistake. If the layer deposition had given the option for layer removal, the layer deposition would have gone backward to remove the layer rather than going forward by depositing on a layer that is a mistakenly deposited layer. Since there is

Fig. 2.6 No connection between next layer and last layer

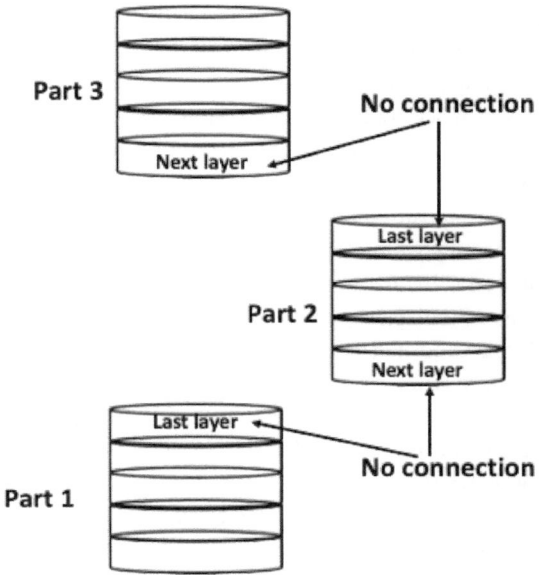

no such option available, layer deposition cannot move backward, i.e., there is no space available for the layer deposition to take place even if the layer deposition intends to move backward. This is the lack of space that says the layer deposition does not move backward because it cannot move backward.

But when part-1 cannot be made bigger because the deposition cannot take place, this inability of the layer deposition is not because there is no space available for the layer deposition to take place. There is already space available. In Fig. 2.6, there is space at the left of part-2 and above part-1. If it is said the space is available, it is because its availability is seen from the fact that whether there is an absence of layer deposition at this place. Since there is an absence of layer deposition, it intends to inform that the space is available for the layer deposition to take place.

In bed process, when the deposition cannot take place to make part-1 bigger, it is not because the past layer deposition has made the space unavailable. It is because the layer deposition is not a one-step occurrence. What is one-step occurrence? In deposition process, when a layer is deposited, the question is only asked whether the deposition has taken place or whether the deposition has not taken place. Because in deposition process, the layer deposition is a one-step occurrence, which means the layer deposition is a material deposition, and vice versa. But in bed process, the layer deposition is not confined to only one step that is the material deposition in the deposition process. It means the material that is deposited, in the bed process, may or may not lead to layer deposition. When it is said that the space is available at the left of part-2 and above part-1, it means the space is available because the layer deposition has not taken place, it does not inform whether the material deposition has taken place. A space will be available for layer deposition to take place, in the bed process, only when neither material nor layer deposition has taken place. In the bed process, when the material deposition has taken place and layer deposition has not taken place, the layer deposition cannot move backward not because the layer deposition intends to un-deposit the past deposited layer but because the presence of the past material deposition does not allow to initiate the layer deposition.

In deposition process, the space is available does not mean there is a presence of material in that space. Thus, if there is space available at the left of part-2 and above part-1, the space can be filled and part-1 can be made bigger. If the part cannot be made bigger and if it is because the geometry of part-2 does not allow to make part-1 bigger, the geometry can be rectified, i.e. the problem can be rectified to make part-1 bigger. In bed process, part-1 cannot be made bigger even if the geometry of part-2 is changed. Thus, the problem that does not allow to make part-1 bigger, in bed process, is not the problem that arises due to the local condition that can be rectified. This problem in bed process is the problem of the process. This is the constraint of the process. As long as the process will be there, this problem will be there. Thus, if it is said layer deposition moves in a forward direction, it has two meanings in a bed process: (1) a deposited layer cannot be undeposited, and (2) if the connection between two layers is lost, a layer cannot be added again to undo the loss, i.e., there is no hope for a part to regrow. While in a deposition process, it has only one meaning, i.e., a deposited layer cannot be undeposited. Thus, when the connection between two layers is lost, the part can regrow when the layer deposition is resumed in the deposition process. It means when the connection between two layers is lost, the part can regrow only in the deposition process.

In bed process, when the next layer cannot be deposited to make part-1 bigger, and the layer deposition starts forming part-3. This is a problem but this problem has an advantage. The advantage is that the position of the layer cannot be between above part-1 and below part-3. It restricts the coordinate the next layer can have. It says the coordinate of the next layer cannot choose any value it wants to choose. This saves from guessing from a number of the positions the layer can take. What is the implication of this restriction? This restriction says that the way a part is formed in bed process is restricted. It means there are not many ways a part can be formed. There are not many ways a design can be converted in a given orientation. It says there is a hope that how a design can be converted will be equal to how the design will be described, i.e., there will be no difference between a conversion and a description. It means it will be better than using coordinates because the coordinates only say where the layers are located, it does not say how the layers got located. It means the method of using the next layer and the last layer will be better than using coordinates that do not yet exist.

But the problem is that this problem does not exist in a deposition process. If the problem is not there, the advantage that is gained from the problem must not be expected as well. The advantage gained from the problem was that there was a restriction in the direction of fabrication in the bed process. The deposition process cannot expect to have a unique fabrication direction. It can move in a backward direction if there is a space. Or it cannot move in a backward direction even if there is a space because there is no order in which direction it will proceed. This brings a variety in fabrications. This brings chaos when a unique fabrication is sought. This brings chaos when a unique description of the fabrication is sought.

Figure 2.7 shows a design whose conversion into a part can be done in only one way in a bed process. Therefore, the fabrication direction is only one. The design changes at point 1 when a layer is fragmented into two parts. There is another change at point 2 when the layer becomes smaller in area only for the second part of the layer. There is no further change except the layer deposition stops after some height. This description of the fabrication and this fabrication direction will also work for deposition process. But the deposition direction can move backward as well in the deposition process, requiring different description for the design.

Figures 2.8, 2.9 and 2.10 show three ways how the design can be converted. Figure 2.8a shows fabrication direction moves forward in step-1. Figure 2.8b shows the connection is lost between the last layer and the next layer as the layer deposition moves backward. Figure 2.8c shows the connection is again lost between the last layer and the next layer, and the layer deposition starts afresh at the right the side of the part. Thus, for converting the design, the layer deposition moves forward, then backward, and then forward. Figure 2.9 shows the conversion of the design takes two steps, while Fig. 2.10 shows the conversion takes three different steps. Fabrication as shown in Fig. 2.7 shows only two changes (1, 2), while fabrications in Figs. 2.8, 2.9 and 2.10 show three changes (1, 2, 3). Fewer changes means fewer sources of problems. Therefore, Fig. 2.7 is preferred as

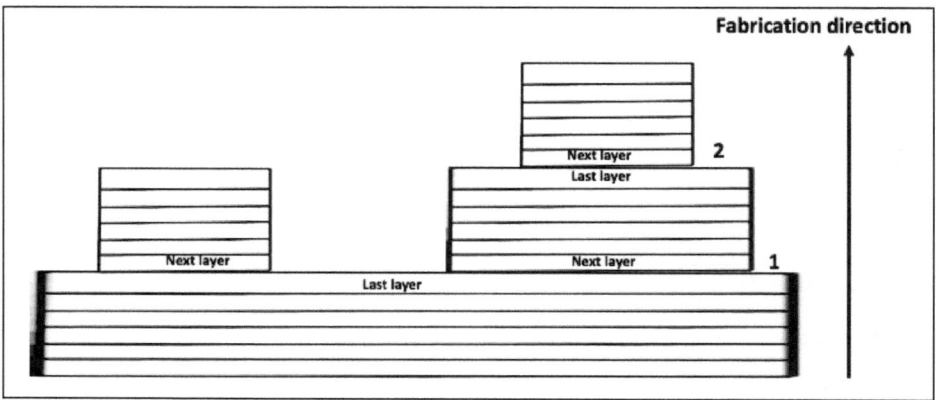

Fig. 2.7 Conversion of design in bed process

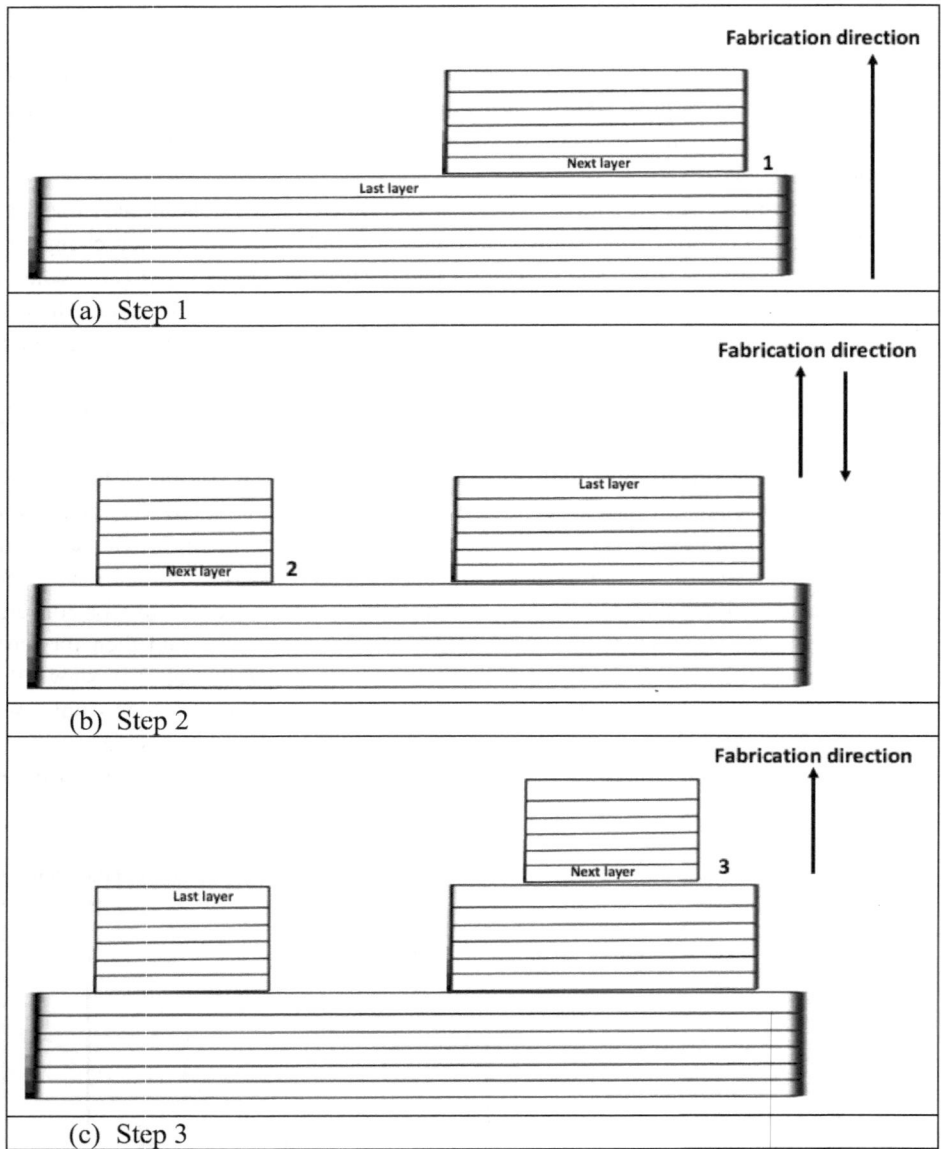

Fig. 2.8 Conversion of design in deposition process: method 1

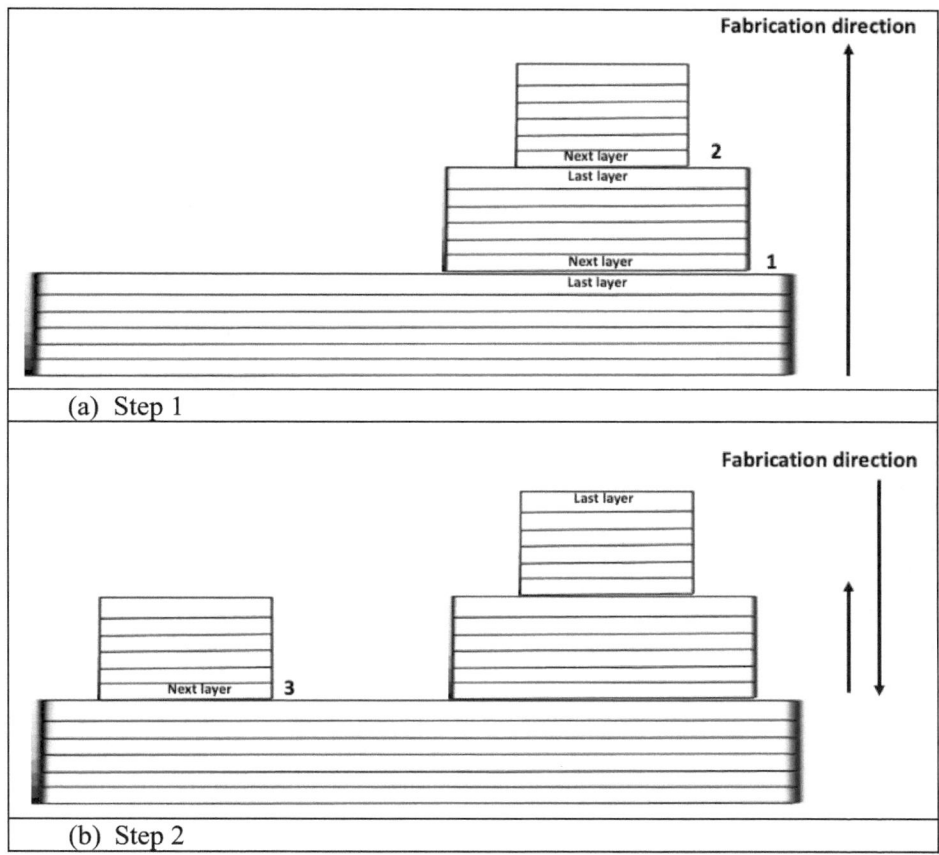

Fig. 2.9 Conversion of design in deposition process: method 2

it will create fewer problems. This also shows bed process means fewer changes while deposition process means fewer changes only when the deposition process follow the steps as the bed process does. This means, in general, bed process is always better than deposition process.

2.7   What Is the Limitation of the Definition of a Layer?

The difference between a connection and no connection is the absence of a layer, even in fragmented form. If the layer is shifted at an extreme position with respect to the last layer, so the connection is little, it will save the situation from becoming the state of no connection. When there is such a change in the layer, this can be explained with one of the

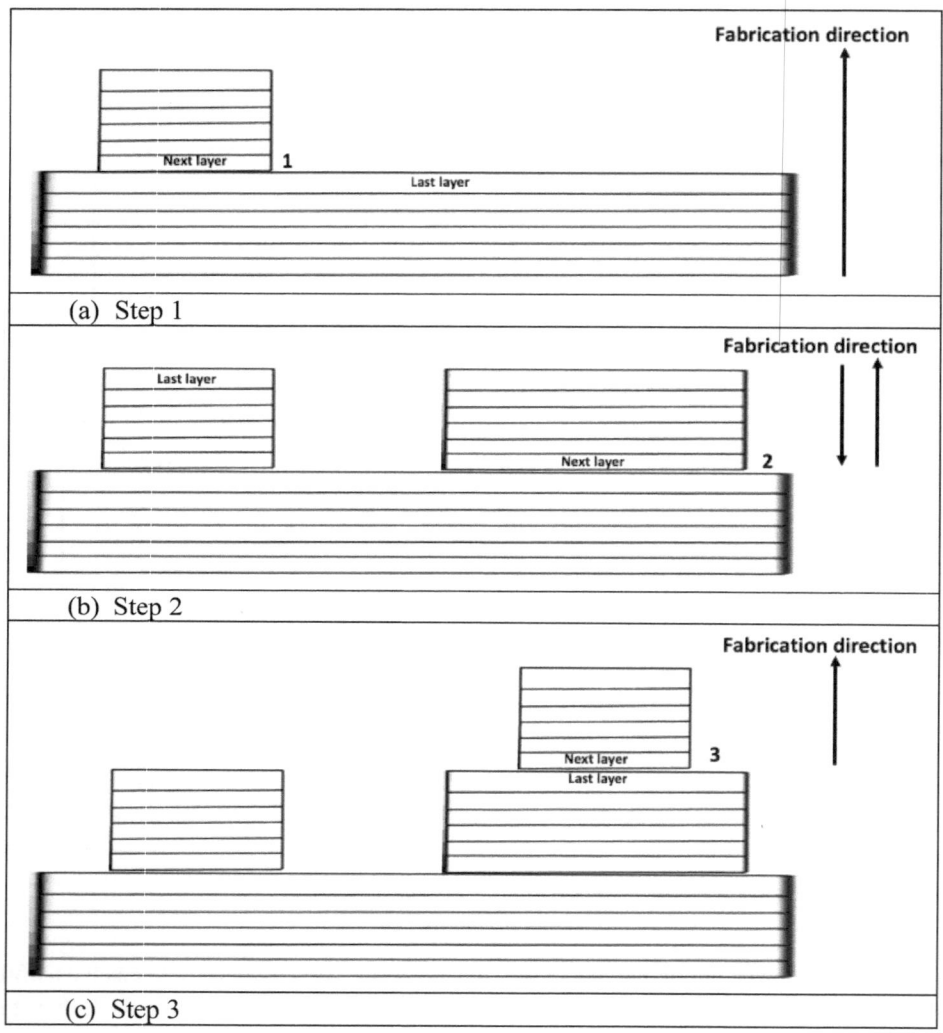

Fig. 2.10 Conversion of design in deposition process: method 3

variable of a layer, i.e. location. As per the definition of a layer, there are four variables: thickness, location, shape, and area. If a layer changes using one or any combination of four variables, this change can be taken into account. But, if a layer is fragmented and still has a connection with the last layer, this change in a layer cannot be taken into account using the four variables. It implies that the definition is not adequate. When a layer is fragmented, the layer does not exist anymore. The definition is for that layer that exists.

Thus, the definition of a layer that is applicable to an individual layer is not applicable when the layer is part of a part.

An individual layer is a layer because there exists a connection from its one side to another side. As long as there is a connection, it does not make any difference how various sections of the layer are connected. If the connection is lost and various entities of the layer are still connected, it will still be a layer. The state of being connected is more important than having a connection, because this state allows a layer to exist as an integrated unit. When a layer is an individual layer, the difference between having a connection and being connected is not clear; both are the same things. When the layer becomes part of a part, the absence of a horizontal connection does not allow it to lose its state of being connected because of the presence of vertical connections. When a layer is defined, it is defined on the basis of the connection it has, i.e. a horizontal connection. It is not defined on the basis of the state of being connected; it is not taken into account that the state of being connected is more important than how this state is achieved. Since AM does not make a part that has only one layer, the definition of a layer needs to be extended. If the definition cannot be extended because the definition is expected to be only for those layers that are still layers when they are individually verified taken out of a part, a layer parameter is required to take into account the fragmentation.

2.8   Is Layer Thickness Required to Define a Layer?

When a design is divided into layers, the basis for the division is that the layers will be fitted well in the design so they will collectively represent it. Thus, a layer represents part of the design. A layer can represent part of the design only when its projection is equal to part of the design. The projection will be equal to part of the design only when part of the design has no contour. Thus, a rectangular block requires one layer that has thickness equal to the height of the block. If a block consists of two small rectangular blocks having different widths, this composite block will require two layers, each layer having a height equal to each small rectangular block. If a block is bent, it will require multiple layers so their projections will try to match part of the block. The number of multiple layers depend upon the angle of bent—if the angle is high, the number will be high. This is how the collection of multiple layers will represent the design of the bent block (Fig. 2.11).

A rectangular block needs to be divided into one layer. A composite block needs to be divided into two layers. The bent block needs to be divided into multiple layers. The division of the blocks is not based on their heights but on the basis of the contour they have. Since the rectangular block has no contour, it does need more than one layer. Therefore, the thickness of the layer of the block can be many depending upon the height of the block, while the number of layers is fixed, i.e. one. Thus, the number of layers comes first, layer thickness comes later. Layer thickness depends on number of layers.

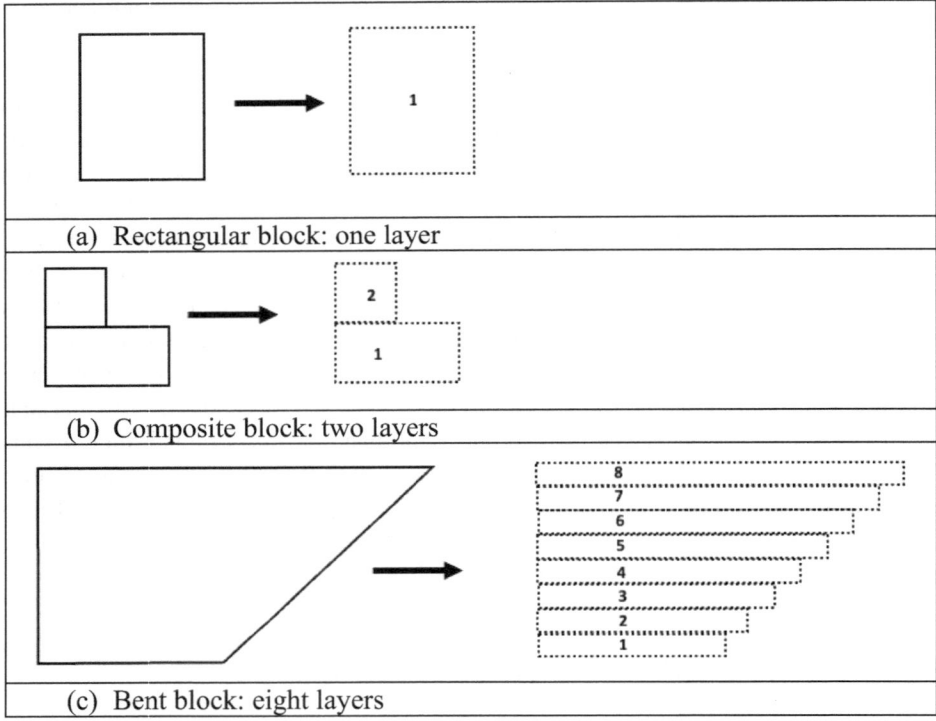

(a) Rectangular block: one layer

(b) Composite block: two layers

(c) Bent block: eight layers

Fig. 2.11 Number of layers required according to projection: **a** one layer, **b** two layers, and **c** eight layers

Therefore, if number of layers is used as a variable, layer thickness as a variable is not required.

This is what comes from projections of layers. But the layer thickness derived from the projections of a layer will not help manufacture a part. For example, a rectangular block cannot be converted from design to a part by considering it one layer. This conversion will depend upon the ability of AM technique even though projection does not prevent it to be converted. If the block cannot be converted because layer thickness is high, layer thickness needs to be decreased. Similarly, the bent block is divided into multiple layers depending upon how the projection will help realize a design. But if the layer thickness is too small to be made, the layer thickness needs to be increased. If the layer thickness is changed to facilitate the conversion, this change is not what is suggested by the design or layer arrangement. Thus, there are two layer thickness: one that comes from projection, and another that comes from manufacturing limitation.

A layer thickness that comes from the projection will not necessarily help manufacture a part, and the layer thickness that comes from manufacturing limitation may not help know futility of layer thickness as a variable for describing layer arrangement. Thus in this case, layer thickness as a variable is required to manufacture a part but is not required to describe layer arrangement. But layer thickness can be known from the number of layers, and vice versa, as layer thickness does not change in a given design.

Thus, when layer thickness does not change in a design, layer thickness is a primary variable, while number of layers is secondary variable to manufacture a part. But the number of layers is a primary variable, while layer thickness is a secondary variable to describe layer arrangement.

In layer arrangement, if the number of layers is selected as a variable in place of layer thickness, how to describe a single layer? A single layer requires a layer thickness to be defined. But a single layer does not exist in AM. AM is an arrangement of layers. However, if a part is made up of single layer, it can still be defined by the number of layers, i.e. one, and its thickness can be known. It does not mean how much the number of layers is a useful variable that can also help define a single layer, it only means what the need is to define a single layer when AM is not about a single layer acting independently.

When layer thickness changes in a design, layer thickness is required to describe layer arrangement. But layer thickness is not an independent variable.

When a design is given, its vertical dimension is fixed. Layer thickness has dependency that it is supposed to fill the vertical dimension of the design, the design is not supposed to adjust its height to be filled well by the layer thickness because the design is given. 'The design is given' is an excuse. It is not because the design must not be respected and be converted into a part, but because the design is not respected in AM. The design is changed, so it will match the layer arrangement. The design is not respected when an approximation is desired. The shape of layer is changed to meet the shape of the design, the area of the layer is changed to meet the shape of the design. But when layer thickness comes in question, it is expected that it should respect the design and fill up the design by adjusting itself. It is because the shape and the area of the layer are independent variables. They do not have dependency on any other variables. But layer thickness is not such an independent variable.

2.9   What Happens When a Layer Is Not Changed in Layer Deposition?

In layer deposition, if it is known how a layer that is going to be deposited will be positioned with respect to the layer on which it is getting deposited, it will say how the resulting layer arrangement will be. Therefore, only two layers, i.e. the next layer and the last layer is required to track changes. In Fig. 2.12, there is no change between the next layer and the last layer, because both layers have the same thickness, shape, area, and horizontal position. The next layer is shown by a red line, while the last layer is shown by a black line. Since there is no change, the top view of the layer arrangement only

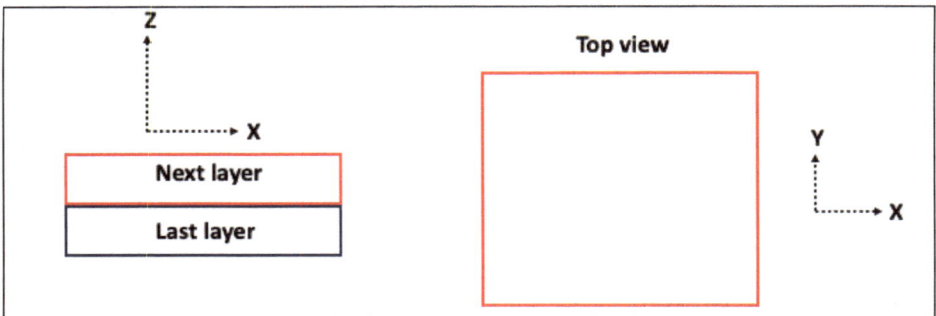

Fig. 2.12 Next layer does not change with respect to last layer

shows the red line, as the black line is hidden by the red line. The height is in z-direction, while the area of the layer is in an xy-plane.

Following are the consequences of no change in a layer:

1. The shape of a product can be known by knowing just one layer. For example, if the layer is a ring, the product will be a hollow cylinder. Since a layer is a 2D entity compared to a product that is a 3D entity, AM reduces the problem of making a 3D product to the problem of making 2D layers. But in this case, the problem is reduced to making a single layer.

   The difference between a product and a layer is the absence of height. The height, which is a third dimension, is not acquired by learning how different layers are made but by repeating the parameters for making a single layer. The pre-knowledge of the height is not required as it does not affect how layers are made. This gives freedom to scale the fabrication after the fabrication starts. This is how in this case, AM is different from other techniques.

   In other techniques, when a product is made, all its dimensions must be known before starting to use the technique. For example, in machining, the size of a product must be known to select the size of a machining tool; in casting, the size of a product must be known to select the size of a mold. But in this case of AM, the height of a product can be an afterthought, i.e., AM, unlike other techniques, gives freedom from the third dimension.

   Although, this freedom has limitations as the height of the cylinder cannot be more than the height of a machine (especially in bed process). This requires the pre-knowledge of the size of the machine to determine the height of the cylinder. But this requirement is not because AM in this case will change with a change in machine height but because AM will not be able to work if the machine height is not sufficient. Therefore, as long as AM is able to work, the lack of the pre-knowledge of the

Fig. 2.13 Net-like layer

height of the cylinder will not prevent it from continuously adding to the growth of the cylinder.

2. Since subsequent layers are made by repeating the same parameters, increasing the number of layers does not increase that manufacturing task that is required when each layer does not need the same parameter. Although, that manufacturing task that is required to make more layers is increased. Therefore, the number of layers can be increased without finding parameters for every layers. Thus, an increase in the manufacturing task with an increase in the number of layers comes almost free. This convenience gives advantage—the design can be divided into an increasing number of layers if increasing the number leads to advantages, such as an increase in accuracy or strength.

   Repeating the same parameters may not work if the height is large enough to cause changes due to gravity or heat accumulation. For example, if a layer is like a net consisting of thin sections, these sections may collapse or may not allow to make another layer on it (Fig. 2.13). Or the layer deposition may cause the heat to accumulate, which will require further parameters to supply less heat. Thus, the same parameters will not work after a critical height. But before the critical height, the parameters that can be repeated can be obtained conveniently.

3. Repeating the same parameters for making layers means an increase in speed from conceptualization to fabrication. This is the advantage in this case of AM, which says a high fabrication speed can be obtained only because fabrication happens layerwise. Though layerwise is known to simplify complexity, this case shows why AM can be used to accelerate fabrication.

4. Parts made are simple. But if these simple parts are combined, the combination will give a complex part. But the problem is how to combine. For example, cylinders and rectangular blocks are simple parts, if they are joined anywhere, i.e., at the middle, bottom, or top, this joining will bring a change in layers. Since in this case, there must be any change in layers after the start of the fabrication, it is not possible to join these simple parts anywhere. Thus, making a complex part is not possible by joining. But if the substrate is included as a section of the final part, a complex part can be formed. However, if the substrate is included, it will not be AM. Because AM does not make a substrate while it makes a part on it.

   Figure 2.14a shows a design consisting of four cylinders (1, 2, 3, 4) of different diameters and two rectangular blocks (A, B) of different widths. All have same heights. This combination of six shapes and a substrate is simple to be made in AM but is complex in machining. In machining, one tool may not work for all diameters and widths, or it cannot move when one shape is to be made in close proximity to another shape.

This complex part will be made only when all shapes are of equal height, so the shape of the layer will not change till all shapes are formed. If some of the shapes have different heights (Fig. 2.14b), they cannot be made because the shape of a layer will change when the height of the shorter shape will reach. Thus, in bed process, Fig. 2.14b cannot be made.

However, in deposition process, Fig. 2.14b can be made only when a shape is made after the fabrication of another shape. When the shapes are made serially, there will be no change in the layer for making a shape. Thus, all shapes can be made as these

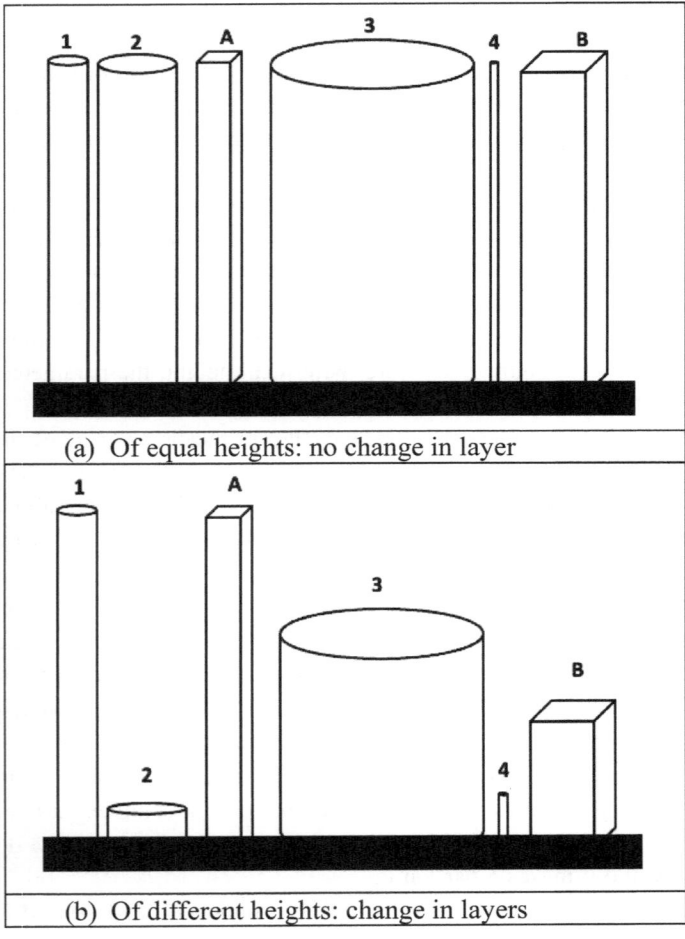

Fig. 2.14 Complex part: **a** no change in layers, **b** change in layers

shapes do not change and therefore, do not require a change in layer to make it. But, if the shapes are made serially, the fabrication of the next shape will be affected by the geometry of the last shape. Therefore, it will not allow to make shapes of all heights. It will then be difficult for it compete with machining. As a machining tool cannot access when the design of one shape is far smaller than that of another shape, a nozzle, for example, will not access as well.

2.10   What Are the Various Possibilities of Layer Arrangements?

If there is a change in layer arrangement, there will be a change in a part that forms. The layer arrangement can change in many ways by changing variables that are as follows: layer thickness, layer area, layer shape, layer location, and layer division (Fig. 2.15).

1. **Layer area**: big, small

The area of a layer changes in only two ways: it becomes bigger (Fig. 2.16) or smaller (Fig. 2.17). In both cases, it forms a gap with the last layer. The gap shows why both layers are not same. A gap is a defect. If the gap is not a defect, it shows only a change. A gap as a defect originates when a change brings an undesirable result that is not required. For example, when a layer tries to match a curve, it changes with respect to the last layer to match the curve. The change brings an undesirable result, a gap that is not required. It is known it is a defect because the corner of two layers are joined by an imaginary line. The corners are joined because the line is the closest approximation to the curve. This is the presence of the imaginary line that says a defect is coming. When the layer changes with respect to the last layer, it is not an indication of the gap. The gap comes because

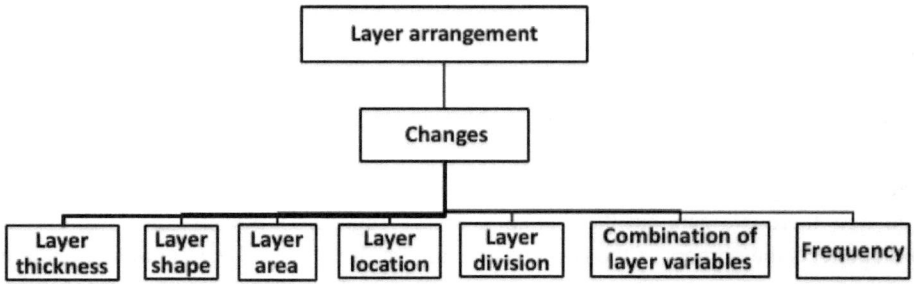

Fig. 2.15 Changes in layer arrangement

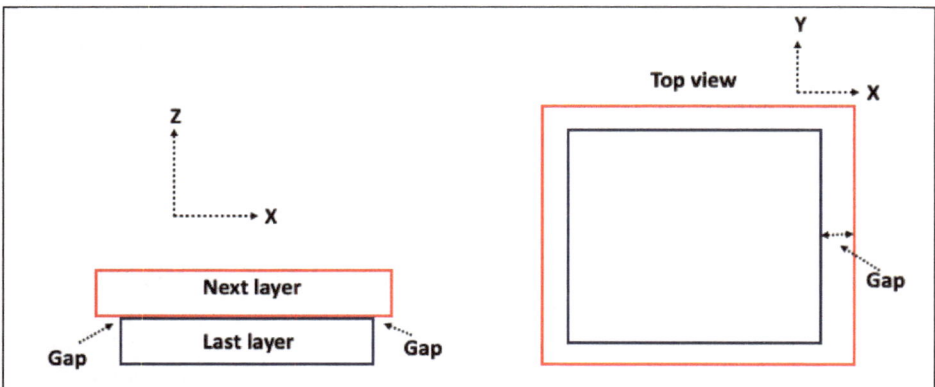

Fig. 2.16 Next layer is bigger

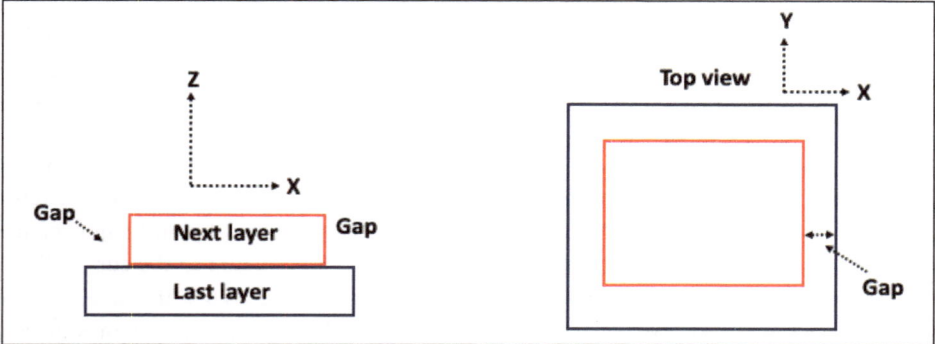

Fig. 2.17 Next layer is smaller

there is an effort to connect two corners. For example, the next layer changes with respect to the last layer; if this last layer is the top layer of a rectangular block and the next layer is the first layer of another rectangular block, then the change between two layers show the change between two blocks. The size of a block changes. When the upper block is made by changing its first layer with respect to the last layer of the lower block, there is a change, there is a gap, but there is not a defect (Fig. 2.18). Because the imaginary line that connects the corners of the two layers is absent. It is absent because this time there is no curve getting formed. The gap is not seen because the design does not draw a line and connects these two corners. The gap is again not seen because this is not what is called a gap in AM. Thus, the presence of a gap does not necessarily imply the presence of a defect.

Fig. 2.18 No gap or defect at point 1 and point 2

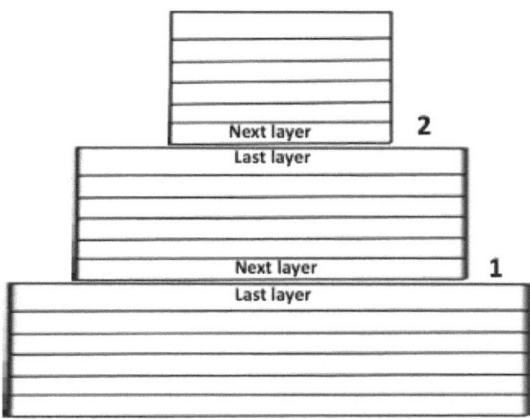

2. **Layer shape**: circular, rectangular, square, elliptical, etc.

If the shape of the next layer becomes different from the last layer, this is a change in shape. Thus, if the shape of the last layer is rectangular and the shape of the next layer becomes circular, this is a change. Since the change does not involve a change in the area, i.e. a change in the planar size, the circle should be closer to the rectangle. It means, for example, the circle should be exactly inscribed in the rectangle, i.e., the circumference of the circle should touch the smaller sides of the rectangle, as seen by the top view (Fig. 2.19). If the last layer were of the shape of a cube, the circle would have inscribed in the square touching all four sides of the square. This is the closer fitting that can happen when the shape change happens without planar size change. This shape change means if the last layer is the top of a rectangular block, the next layer will be the first layer of a cylindrical block (Fig. 2.20). When the circle does not fit in a rectangle, or when the circle does not fit in a square, the size of the corner becomes smaller, this creates a gap. It means the change in shape triggers a change in size, which creates a gap. Thus, a gap is created only when the size change happens. It does not matter how the size change is brought: either by making the layer smaller or by making the layer circular.

3. **Layer location**: the position of a layer with respect to other layers

The location of a layer does not change with respect to the last layer only when there is no change on the layer with respect to the layer. It means the layer is deposited exactly on the last layer. This is one extreme where there is no shift. What will be the other extreme? The other extreme is when the next layer is deposited far away from the last layer so there is no layer deposition on the last layer. Any other positions that the next layer gets are between these two extremes.

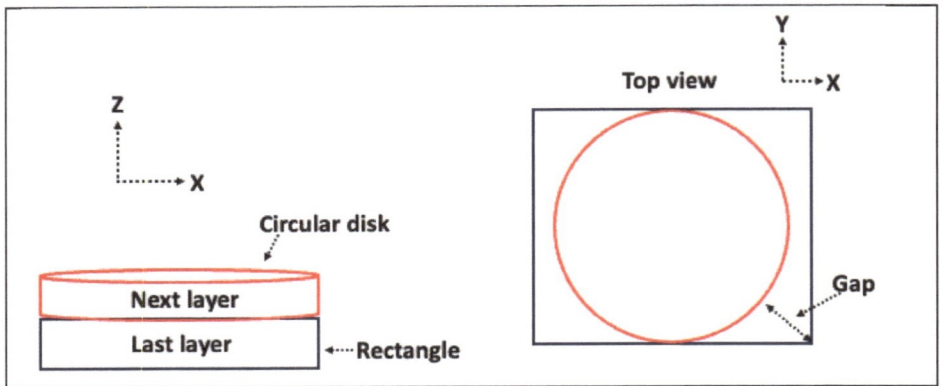

Fig. 2.19 Shape change

Fig. 2.20 Cylindrical block on
rectangular block

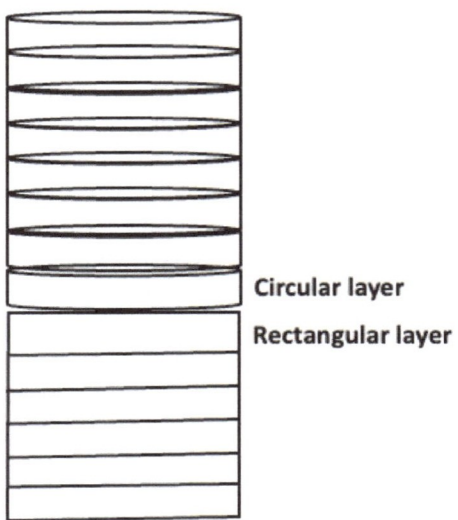

When the layer shifts, it can shift either left (Fig. 2.21) or right (Fig. 2.22), or it can shift either forward (Fig. 2.23) or backward (Fig. 2.24). These shifts can be denoted by shifts in $\pm x$ direction, or $\pm y$ direction in an xy plane. Or the layer can shift in myriad positions that will be made by rotating the next layer at various angles with respect to any of the axes in the plane. Any shift means the layer deposition is moving out of the comfort zone. An outside shift means the next layer has to find a base on which it will be deposited, because the last layer will not completely provide a base, i.e., when the next layer is deposited outside the periphery of the last layer, the problem arises because the next layer is not completely supported by the last layer.

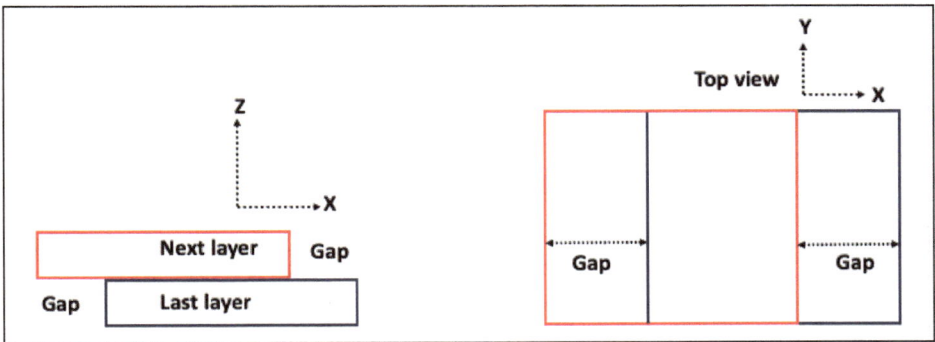

Fig. 2.21   Leftward shift

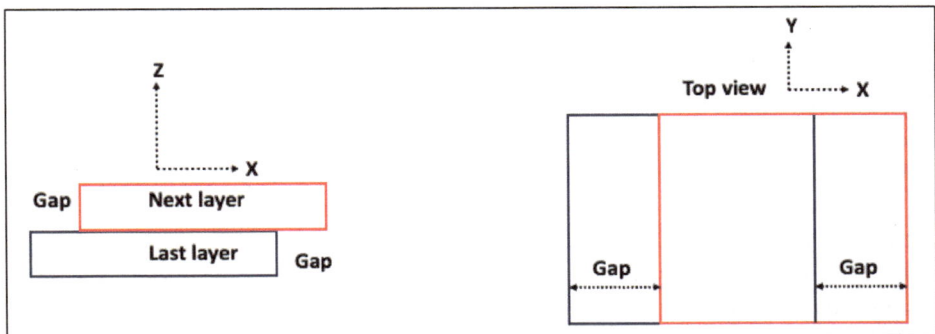

Fig. 2.22   Rightward shift

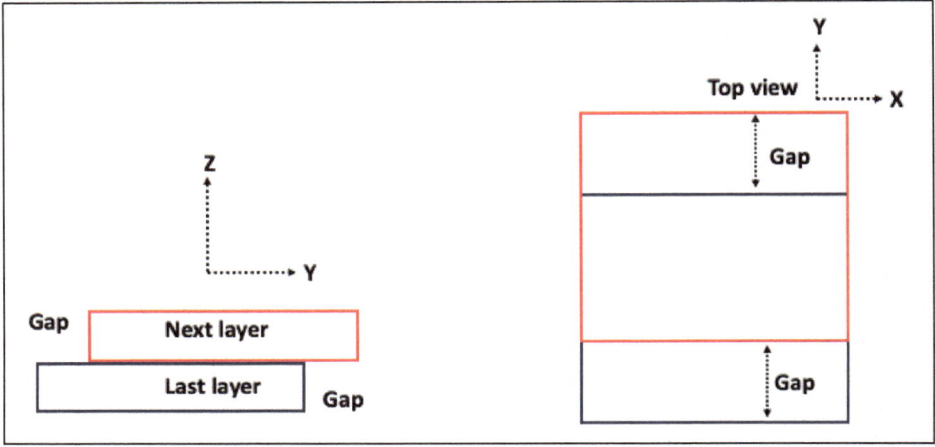

Fig. 2.23   Forward shift

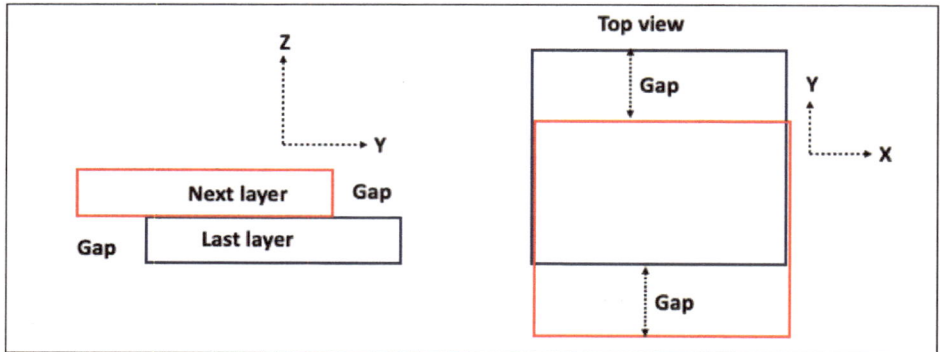

Fig. 2.24 Backward shift

An outside shift means if it is a powder-based bed process, part of the layer is fabricated on the last solidified layer while other part of the layer is fabricated on only powder. For a photopolymer-based bed process, other part of the layer is fabricated on only photopolymer. For a deposition process, other part of the layer is fabricated on air. If it is tried to be fabricated on air, it will fall through. Therefore, either support structure is required to prevent it from falling or part of the layer fabricated on air should be small. Because if it is small, inherent stiffness during fabrication will prevent it from falling.

It means depending on the degree of shift, the next layer may need to find a support structure that needs to be pre-deposited to act as a base. Depending upon the number of shifts, the problem can happen in both directions: one direction to which the shifting happens and another direction from which the shifting happens.

When there are more than one changes in the layers, shift in different directions will give different results. For a single change, due to the symmetry in shape, the same part will form irrespective of the directions it shifts. For an irregular shape, shift in different directions give different parts.

4. **Layer division**: replacement of a layer with various smaller layers

If there is a single layer, its division is not one of its variables. Because the moment it is divided, it is no longer a single layer, it becomes fragments of various shapes and smaller areas. When the disconnected layer is deposited on an ongoing part, it does not fall but remains part of a layer arrangement. Thus, the division as a parameter is required to describe a change that cannot be described by using any of layer variables: area, shape, and location. For example, when the top layer of a rectangular block is replaced with two small rectangular layers to give rise to two rectangular blocks, it cannot be explained using any of the layer variables (Fig. 2.25).

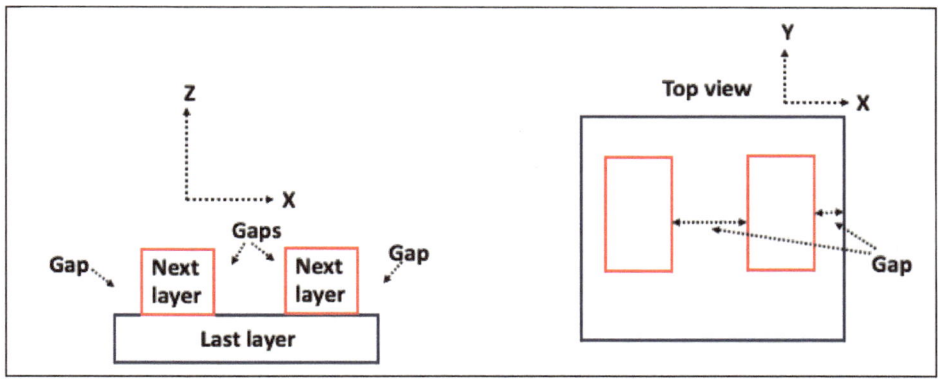

Fig. 2.25 Division

5. **Combination of layer variables**: combination of changes in area, shape, location, and division

When a layer becomes smaller to make a smaller cylinder on a big cylinder, for example. The question arises of where exactly on the cylinder it is located. Whether it is located at the center so that its central axis is coinciding with the central axis of the big cylinder; or whether it is located overstepping the edge of the big cylinder as if it is about to hang, requiring a support. This difference in positions cannot be described using one of the variables, i.e., area, or location; their combination is required. Various combinations give various layer arrangements.

Figure 2.26 shows division and shift of the next layer. Figure 2.27 shows change of shape and shifting of the next layer. Figure 2.28 shows change of shape, area, and location. Figure 2.29 shows the next layer is bigger. When the layer becomes only bigger, there will be shift on both directions of the layer. But this layer is shifted so there is no gap at the left side, while there is a bigger gap at the right side. Figure 2.30 shows shape becomes smaller and the layer is shifted. Figure 2.31 shows the shape of the next layer is different, it is circular while the last layer is rectangular. Besides, the next layer is divided, area changed and shifted.

6. **Frequency**: the number of times a change occurs

When a layer shifts once, it does not create any manufacturing problems that cannot be solved. But this shift is essential because this initiates to make new features, blocks, or sub-parts on an existing part. Figure 2.32 shows only one shift. When there happens two consecutive shifts in the same direction (Fig. 2.33), it starts to form a curve. When only two gaps form, the curve that results will be very small. It is not the practice to form

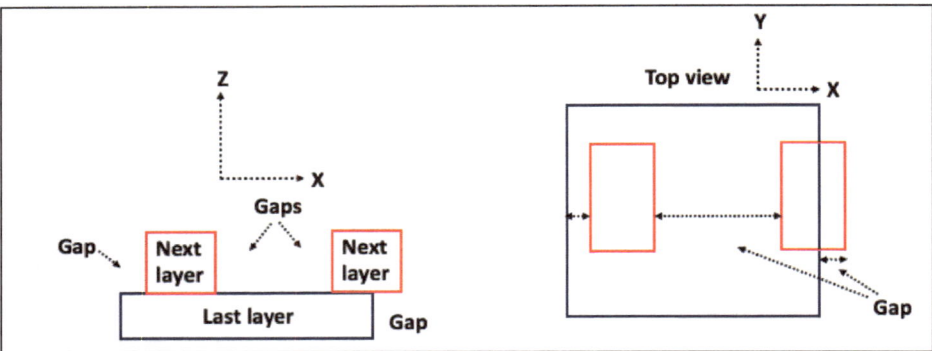

Fig. 2.26 Division and shift

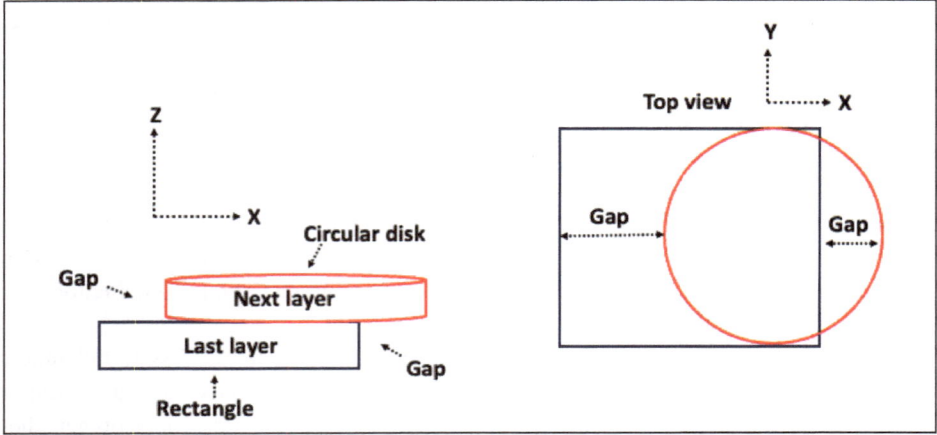

Fig. 2.27 Shape and location change

a curve using only two gaps. But when two gaps are joined, it does not form a normal straight line but a bent or curved line. Thus, with two consecutive gaps, a curve starts to form. These number of gaps is a boundary line to divide fabrication into two types. It means during fabrication in AM, only two things happen: either a curve forms or a curve does not form.

Shifting a layer is not necessary to create consecutive gaps. They can also be created if the deposited layers continuously become smaller (Fig. 2.34) or bigger (Fig. 2.35). If the layer shifts in different direction, the consecutive gaps will not form (Fig. 2.36). The difference between consecutive and non-consecutive shifts is shown in Fig. 2.37.

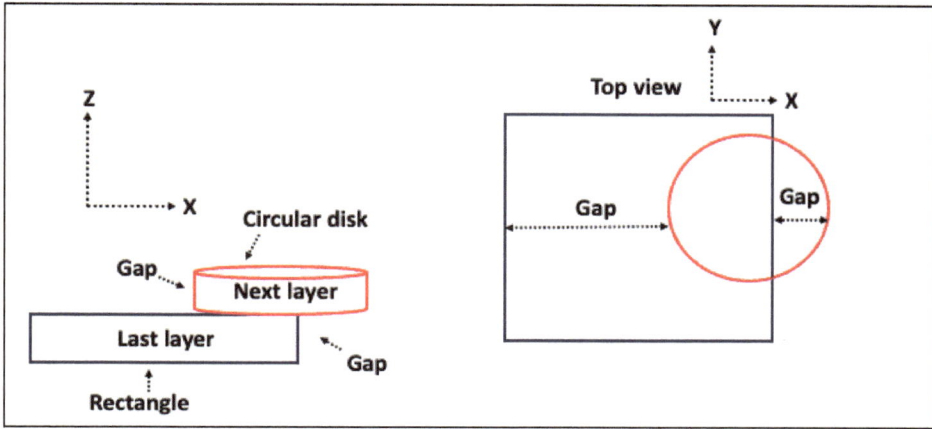

Fig. 2.28 Shape, area, and location change

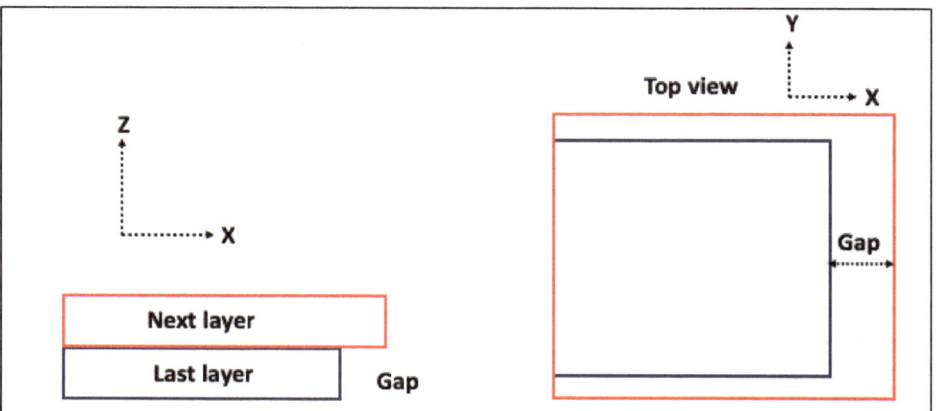

Fig. 2.29 Area and location change

2.11 In How Many Types Can Layer Arrangements Be Divided?

Layer arrangements in AM can be divided into two types (Fig. 2.38):

(1) First type: consecutive shifts in the same direction, i.e., the next layer shifts twice or more in the same direction, and

(2) Second type: no consecutive shifts in the same direction, i.e., except the first type, and

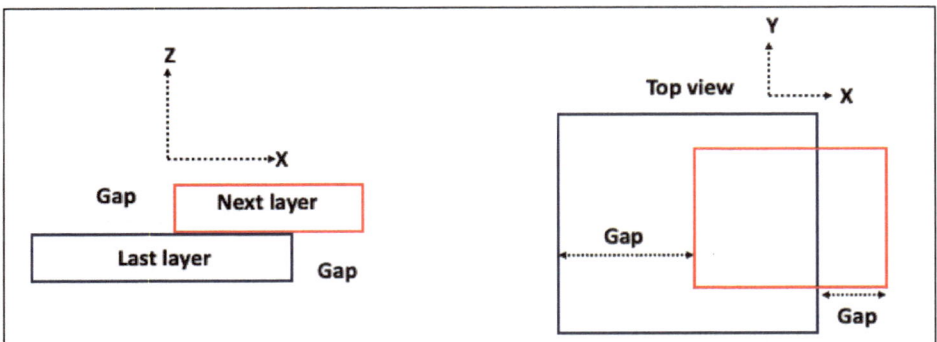

Fig. 2.30 Next layer is smaller and changes location

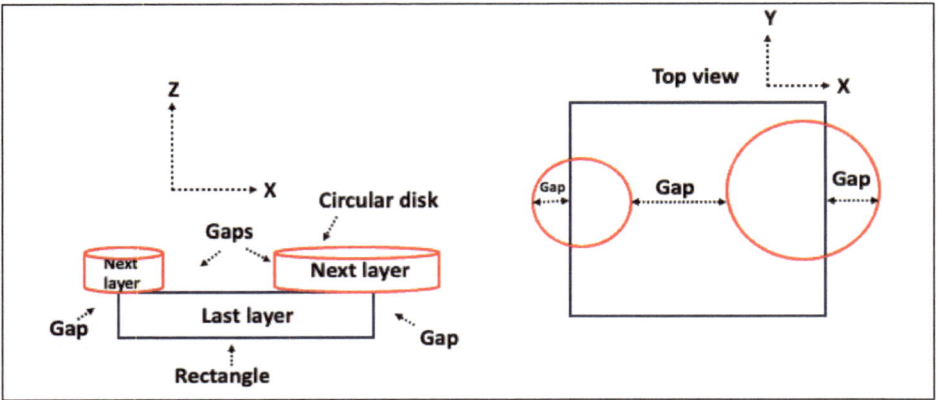

Fig. 2.31 Shape, area, location change and division

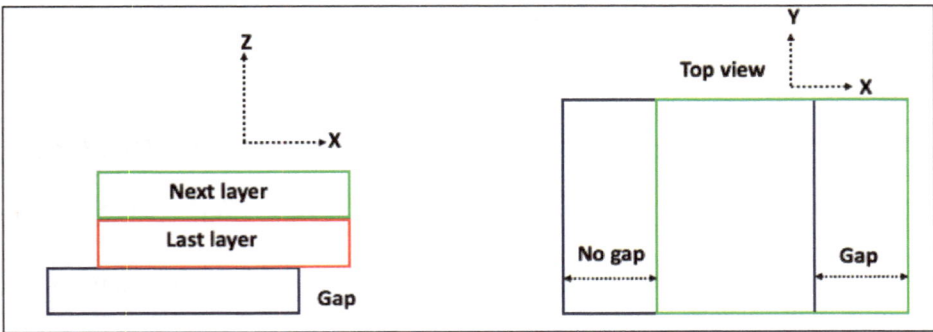

Fig. 2.32 No consecutive shifts

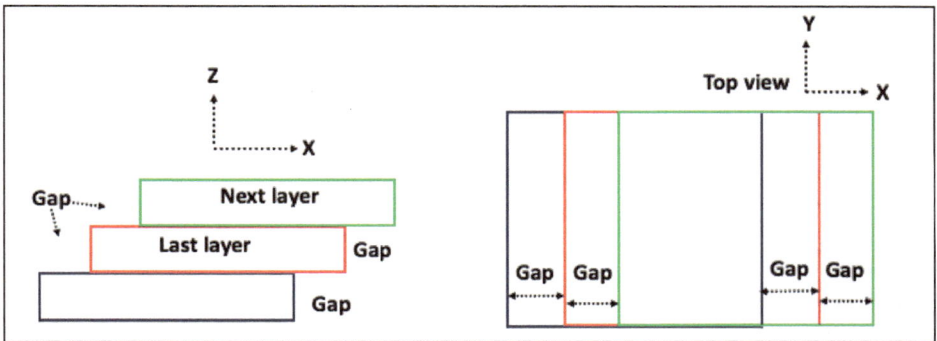

Fig. 2.33 Consecutive shifts in same direction

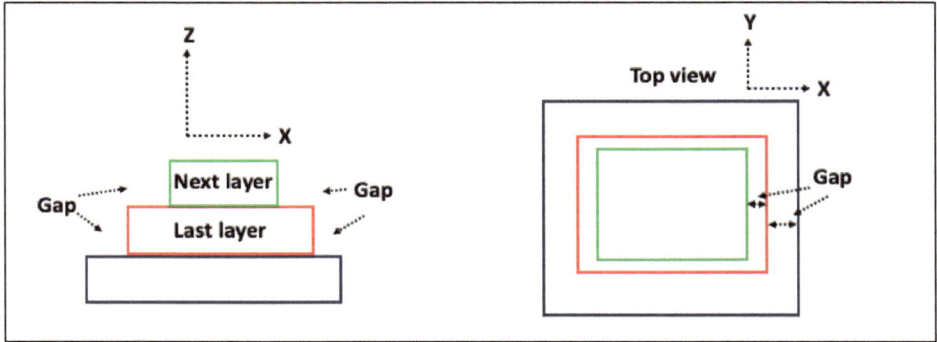

Fig. 2.34 Consecutive shifts in same direction for layer becomes smaller

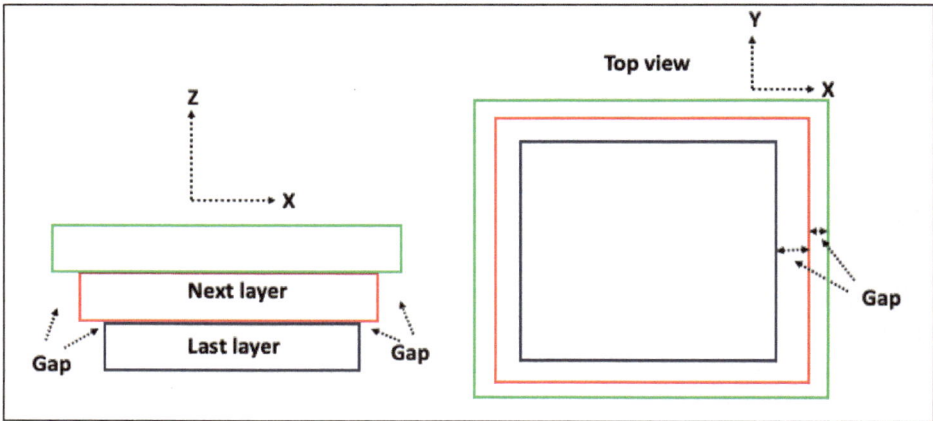

Fig. 2.35 Consecutive shifts in same direction for layer becoming bigger

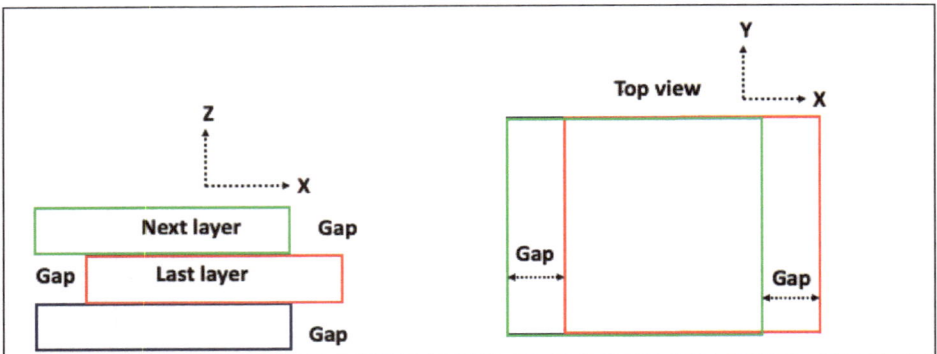

Fig. 2.36 No consecutive shifts in same direction

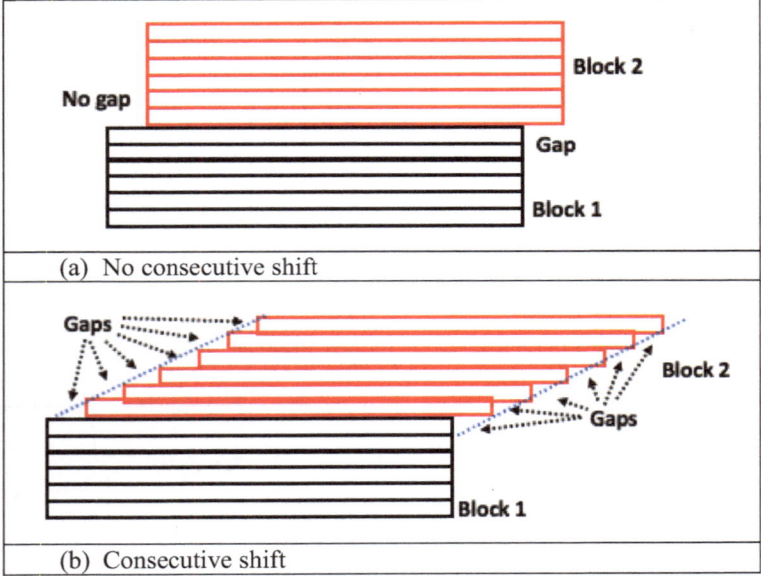

Fig. 2.37 Consecutive and no consecutive shift

Examples of the first type are a cylinder in horizontal orientation, a cylinder of increasing or decreasing diameter, a circular hole on a vertical surface, an arch bridge, a sphere, etc. All other examples that do not fit in the first type are of the second type, such as a cylinder in vertical orientation, a rectangular hole on a vertical surface, a horizontal bridge, a square block, etc.

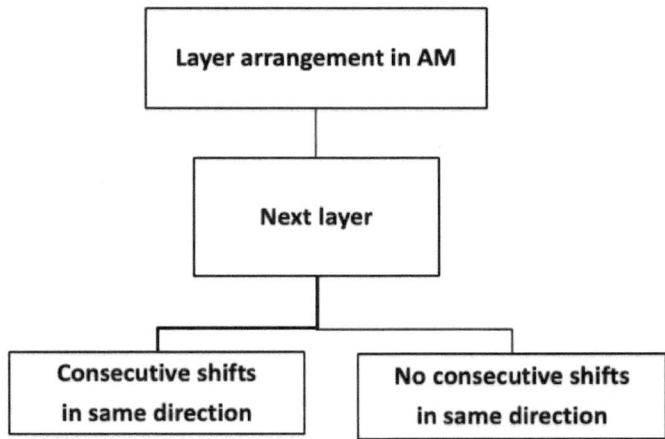

Fig. 2.38 Types of layer arrangement

2.12   What Is a New Theory?

When two or more consecutive gaps are created by shifting of layers during layer deposition, these gaps cannot be filled up, making the problem unsolvable.

Following are the salient points:

1. Change in the deposited layer with respect to the last layer is required to make various shapes.
2. The origin of a manufacturing problem is the change in the deposited layer with respect to the last layer.
3. When the deposited layer shifts outside the periphery of the last layer, it causes more manufacturing problem than when it shifts inside.
4. Manufacturing problems can be divided into two types: unsolvable based on consecutive gaps and solvable based on no consecutive gaps. The problem is unsolvable because the problem lies in the inability of AM to create curvatures.
5. Layer arrangement can be divided into two types: the first type based on unsolvable manufacturing problems, and the second type based on solvable manufacturing problems.
6. The layer arrangement that gives solvable manufacturing problems makes accurate parts.

2.13   What Is the Difference Between Fabrication and Manufacturing?

Fabrication and manufacturing are used interchangeably. AM is not expected to rely on post-processing to improve a part. Therefore, fabricating a part is not different from manufacturing a product.

2.14   What Is Stair-Stepping Effect? How Is It Related to Overhangs?

Stair-stepping effect occurs when the vertical boundary of a layer does not match the contour of a design. Due to the mismatch, there remains a gap between the vertical boundary and the design. When the next layer is deposited, it, like the last layer, again does not match the contour, creating another gap. If these gaps, made by the consecutive layers, are seen, they look like the steps of a stair, giving the name 'stair-stepping effect'.

When the next layer shifts with respect to the last layer to match the contour, the shift creates a gap. Depending on the contour, the direction of the shift changes. When the contour of a design moves away from the next layer, the shift will be outwards to reach the contour. When the contour moves towards the next layer, the shift will be inwards so not to reach across the contour.

There can be two possibilities for the next layer to shift (Fig. 2.39):

(1) Inwards, so there will be a positive gap, and
(2) Outwards, so there will be a negative gap.

When the next layer shifts inwards, the gap is a right-angled triangle, which is one of the steps of a stair. When the next layer shifts outwards, the gap is still a right-angled

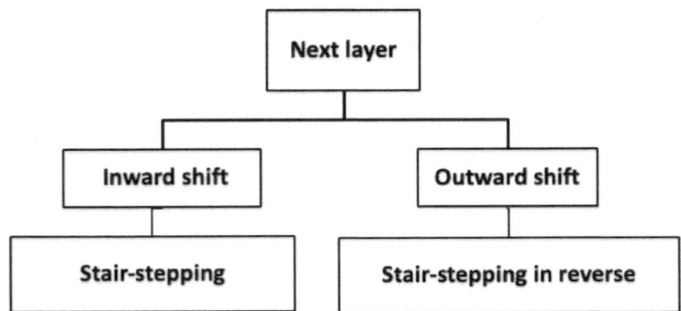

Fig. 2.39 Shifts of next layer

triangle but reversed by 180°, which is similar to the step of a stair reversed by 180°. If the former shift gives a stair-stepping effect, the shift given by the latter must be called 'stair-stepping effect in reverse', which is, though, called overhang.

Which types of contour gives only positive gaps? Or when the next layer shifts inwards, how far should it shift inwards to create only a positive gap? The outcome depends on its size and position with respect to the last layer. When it is smaller than the last layer, it does not need to shift either left or right to create a positive gap. The decrease in its area causes a shift in the position at either sides of the last layer, e.g., a cylinder having constantly decreasing diameter.

In a cylinder having constantly decreasing diameter (Fig. 2.40a), the position of a layer does not shift with respect to the central axis, but the position of the vertical boundary of the next layer shifts with respect to that of the last layer due to the decreasing diameter. Since the shift happens only inwards at either of sides of the last layer, the shift creates a positive gap, a stair-stepping effect on either sides. As long as the position of the next layer does not shift with respect to central axis, there will be stair-stepping effects on either sides.

The problem starts for a contour, where the next layer starts to shift with respect to the central axis. When it shifts rightwards, there will be more stair-stepping at the left side than at the right side (Fig. 2.40b), and vice versa (Fig. 2.40c). The cylinder then is no longer of diameter that decreases uniformly from all sides but is of its left side more bent than the right side. Its shift has a limitation, i.e. the bending of the contour has a limitation. The shift must not cause it to overstep the boundary of the last layer at the right side. If it reaches the boundary of the last layer, the vertical boundaries of both the layers are aligned, giving no stair-stepping effect at the right side (Fig. 2.41a). If the same thing happens leftwards, there will be no stair-stepping effect at the left side (Fig. 2.41b). No stair-stepping effect means no inward shift, which means there is an inward shift at one side and no inward shift at the other side of the layer.

What happens if the shift oversteps the boundary of the last layer? Then it is no longer an inward shift but an outward shift. The outward shift does not only mean a negative

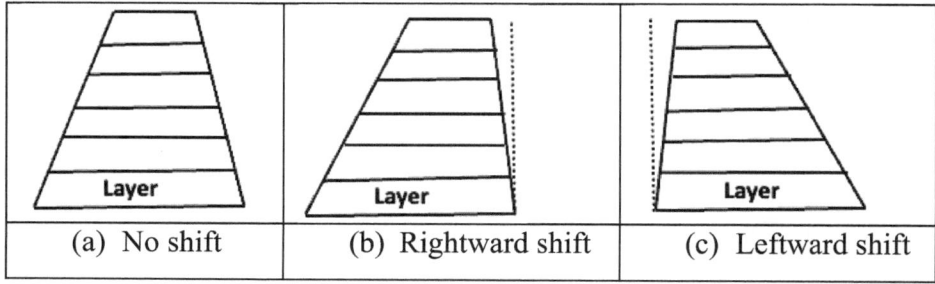

| (a) No shift | (b) Rightward shift | (c) Leftward shift |

Fig. 2.40 Stair-stepping on both sides due to decreasing area of the next layer

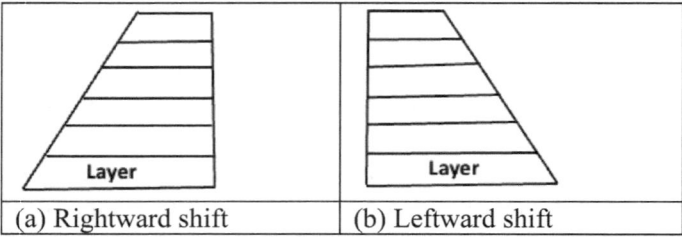

(a) Rightward shift	(b) Leftward shift

Fig. 2.41 Stair-stepping on one side due to decreasing area of next layer

connotation of the inward shift. It means when the next layer oversteps, it brings the question of where to step. As long as there is an inward shift, the next layer has the support of the last layer. But with an outward shift, the next layer needs to be formed on the last layer that does not provide a complete support. With only an inward shift, the formation of the next layer may not require a change in the formation parameter because the next layer does not lack the last layer as a support. But with an outward shift, the formation of the next layer may require a change in the formation parameter.

When the next layer is smaller than (Fig. 2.42), equal to (Fig. 2.43) or bigger than (Fig. 2.44) the last layer, an outward shift at one side means an inward shift at the other side. It means creating an overhang at one side means automatically creating a stair-step at another side. Creating an overhang at one side will not mean automatically creating a stair-step at another side only when the next layer is bigger than the last layer and the size difference between the two layers must be equal to the inward shift (Fig. 2.45).

When the next layer is bigger than the last layer, there will be overhang on both sides without the need to create overhang by shifting the next layer outwards, e.g., a cylinder with an increasing diameter (Fig. 2.46a). When the next layer shifts, there will be more overhang on one side than on another side (Fig. 2.46b, c).

Stair-stepping effect requires to have the steps of a stair. The steps can be made only when the curve is forward-bent rather than backward-bent. Without bending forward, it

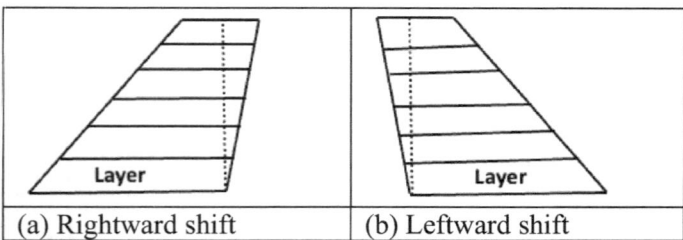

(a) Rightward shift	(b) Leftward shift

Fig. 2.42 Stair-stepping on one side and overhang on another side due to decreasing area of next layer

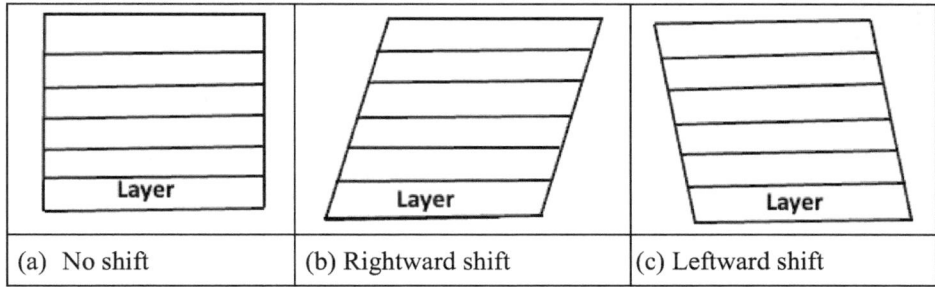

Fig. 2.43 Equal layers: **a** no stair-stepping and no overhang, **b** stair-stepping on left side and over-hang on right side, and **c** overhang on left side and stair-stepping on right side

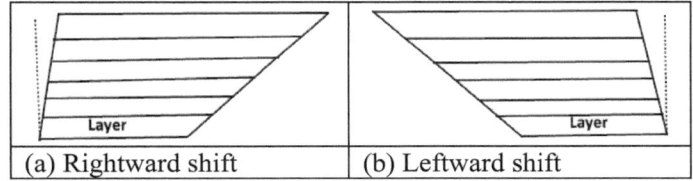

Fig. 2.44 Stair-stepping on one side and overhang on another side due to increasing area of next layer

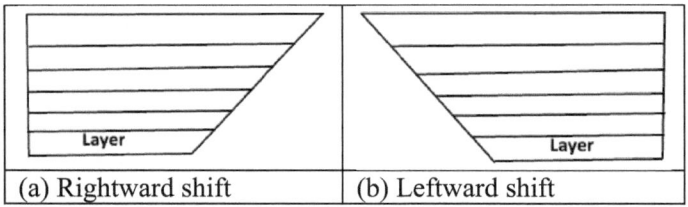

Fig. 2.45 Overhang on one side due to increasing area of next layer

is not possible climb using stairs. Therefore, all those curves that are forward-bent give stair-stepping effects, while those curves that are backward-bent give overhang problems.

Stair-stepping is reversed to overhang, which can be checked by reversing a design (Fig. 2.47a) having both stair-stepping (Fig. 2.47b) and overhang (Fig. 2.47c). Figure 2.47b, c each show the one half of a design vertically divided into two halves. When the design is reversed by 180°, the stair-stepping becomes overhang (Fig. 2.47d), and vice versa (Fig. 2.47e).

Fig. 2.46 Overhang on both
sides due to increasing area of
next layer

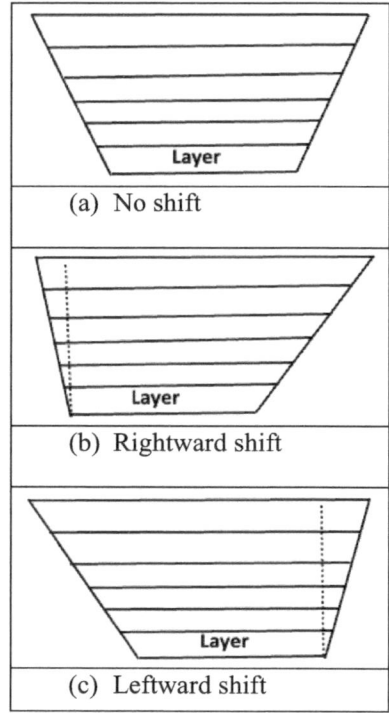

(a) No shift

(b) Rightward shift

(c) Leftward shift

2.15   Why Is Stair-Stepping an Unsolvable Problem?

When an upright cylinder is deformed by tilting it sideways while keeping its base fixed, the shape of its layers changes from a circular disk (Fig. 2.48a) to an elliptical disk (Fig. 2.48b). The elliptical disk does not have an upright but a tilted boundary. This is a problem. Because all layers in AM have an upright boundary, which do not allow their boundaries to match the required tilted boundary (Fig. 2.49).

Why cannot an elliptical disk with a tilted boundary be created? Because layer formation is not completely an artificial process. For example, it can be checked with bed process how a layer forms.

In bed process, when material is pushed sideways to create a bed, the creation does not entirely depend on how the bed is pushed but also on how the material is settled afterwards. Pushing the material is an artificial component of the process, while the material being settled is a natural component of the process. How the material pushed can be controlled but how the material settled cannot be controlled. Because if material settling is controlled, it cannot be said that the material is allowed to settle. Depending upon the device used to push the material sideways, the material can be pushed downward as well. Then the material is deprived of some chances to go downward on its own and settle,

Fig. 2.47 Reversing design by
180°: **a** design,
b stair-stepping, **c** overhang,
d reversing figure (**b**), and
e reversing figure (**c**)

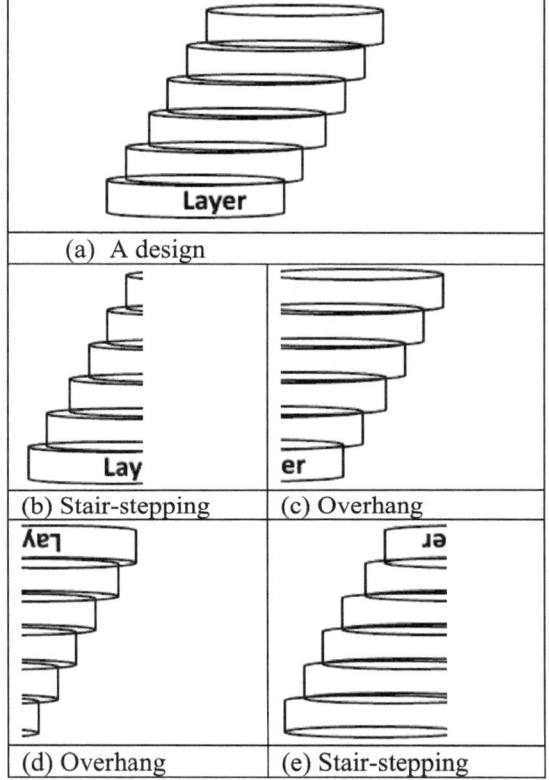

which decreases the amount of material that settles and becomes part of the bed. If the
bed is pressed afterwards, which is not a general practice, it will decrease the influence
of settling on bed density. But it will not be able to overcome the influence of material
settling on the contour of the bed. Though, the contour of a bed is neither of any use nor
important, the contour is mentioned because when material flows to make a bed, the flow
has a pattern that can be noticed when the contour of the flow, i.e., of the bed is checked.
Therefore, whether material is pressed or not, the formation of the bed is not free from
the natural component of the process.

A bed is not formed to have a shape that will contribute to the shape of a product that
is formed on the bed. This simplifies the formation of the bed. This is also the reason
why its formation is not completely controlled by removing the natural component of
its formation. This distinguishes the processing of material to form a bed in AM from
the processing of materials to form other shapes in non-conventional techniques, such as
sintering.

If the formation of a bed cannot be completely controlled, whatever formed on the
bed cannot be completely controlled. For example, when a liquid binder is dropped on a

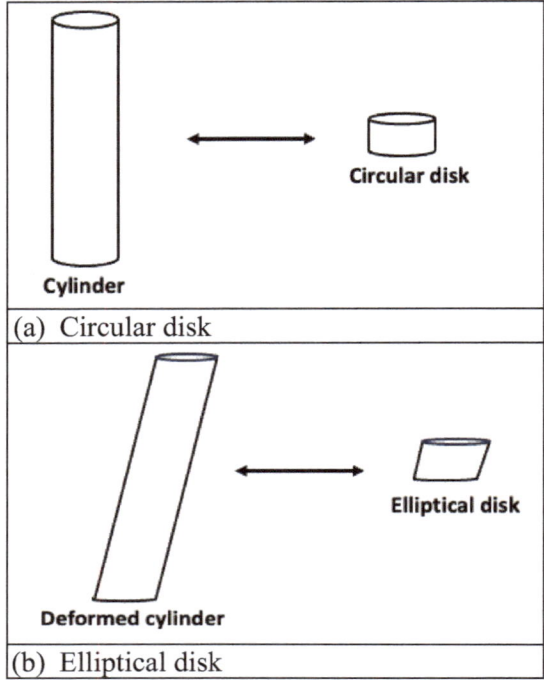

Fig. 2.48 Layer shape: **a** of cylinder, and **b** of deformed cylinder

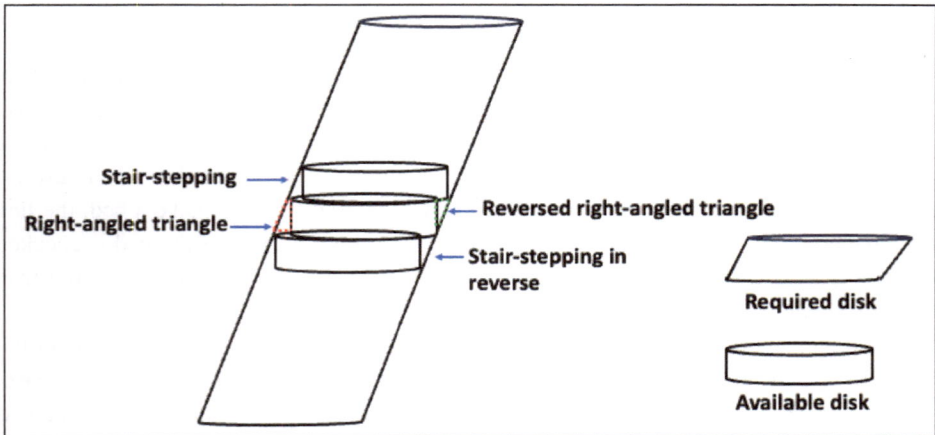

Fig. 2.49 Fitting of circular disks in deformed cylinder

porous powder bed, the binder will mostly flow downwards through porous channels. It will result in consolidation, normal to the bed, i.e. if the binder is dropped along a line on the bed, it will result in a vertical wall. Suppose the formation of the bed is controlled, so it is densified. If the binder is dropped on a dense bed, the binder will not go downwards as far as it went in a porous bed; the binder will also have time to flow sideways. If the binder is dropped along a line, it will not result in a vertical wall but a wedge-shaped wall having wider upper end.

The shape that is created on a bed depends on the properties of the bed. The property of the bed depends upon how much the formation of the bed is controlled. If the formation of the bed is not controlled resulting in a porous bed, it is not possible to be free from the consequence of the lack of control on the wall that is formed on the bed. It is not possible to make a wedge-shaped wall instead of a vertical wall only by controlling the jetted binder. Lack of control means reliance on natural component of the process. When the binder is jetted on the bed, the flow of the binder within the bed relies on gravity that is again the natural component. When the binder is jetted, the purpose of the jetting is to enable the binder to reach the bed, the purpose is not to control the flow of binder within the bed so the binder will travel within the bed as per the contour of the design. If this were the case, there was no need to accept a vertical wall but to make a tilted wall that would have matched the contour, giving no stair-stepping effect.

Controlling the flow of the binder within the bed is difficult, not controlling is easy. Controlling the density of the bed is difficult, not controlling is easy. In AM, the options that are easiest are chosen. This is simplification, which hides the absence of control. This simplification gives rise to stair-stepping effect.

What if the easiest options are not chosen? It means the density of the powder bed is controlled, which will require a device to control the density. A device that is used to control the density of powder is called mold as well. It means going backward in time to re-adopt those devices that are left back to get freedom from them. AM will then lack the benefits that come with the absence of the mold. If jetting and binder are controlled so the binder will flow according to the contour, it will require time for the binding to take place. AM will then not be a fast technique. Thus, to solve stair-stepping problem, AM must be practiced differently than how it is practiced presently without the advantages it provides. It implies that AM will no longer be AM, which makes stair-stepping an unsolvable problem.

However, if there comes a technique that densifies a bed without using any physical device or mold type device, or there comes smart powder that gives the required density, then AM will not lose its advantage that comes with the absence of a mold. If there comes some smart binder and technique that lets the binder move within the bed as per the contour and the binding does not take much time (equivalent to the layer-spreading time), AM solves stair-stepping problem without becoming slow.

2.16   What Is Overhang?

There are two types of overhangs in AM (Fig. 2.50):

1. First type: an angled surface—due to stair-stepping in reverse. Example: a cylinder of increasing diameter at one side (Fig. 2.51). The overhang makes an angle with a substrate. This type is due to the first type of layer arrangement.
2. Second type: a horizontal surface—due to a single layer. Example: a design made up of two upright cylinders, where one cylinder is not placed exactly on another cylinder. The horizontal displacement of the top cylinder with respect to another cylinder gives rise to an overhang that is a horizontal surface (Fig. 2.52a). The surface is parallel to the substrate, i.e. it does not make an angle. This type is due to the second type of layer arrangement.

Figure 2.51a shows an overhang that is made because layers are shifted outside a dotted line, i.e., outside the perimeter of each last layer. It is made because a number of layers contribute to form it by shifting. Its profile depends on which layers shift how much. The shift has made a planar overhang, but it also makes other surfaces: convex (Fig. 2.51b), or concave (Fig. 2.51c). At no point on these surfaces, the next layers shift inwards. If they shift, they will not make overhang but stair-steps.

Figure 2.52a shows that a cylinder is made on another cylinder by shifting the position of the former on the latter. When the next layer of the first cylinder shifts outside the

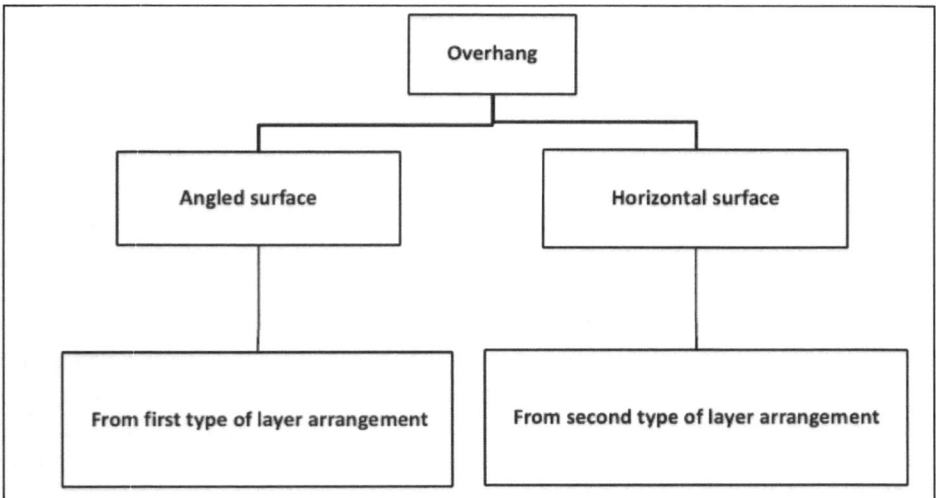

Fig. 2.50 Types of overhangs

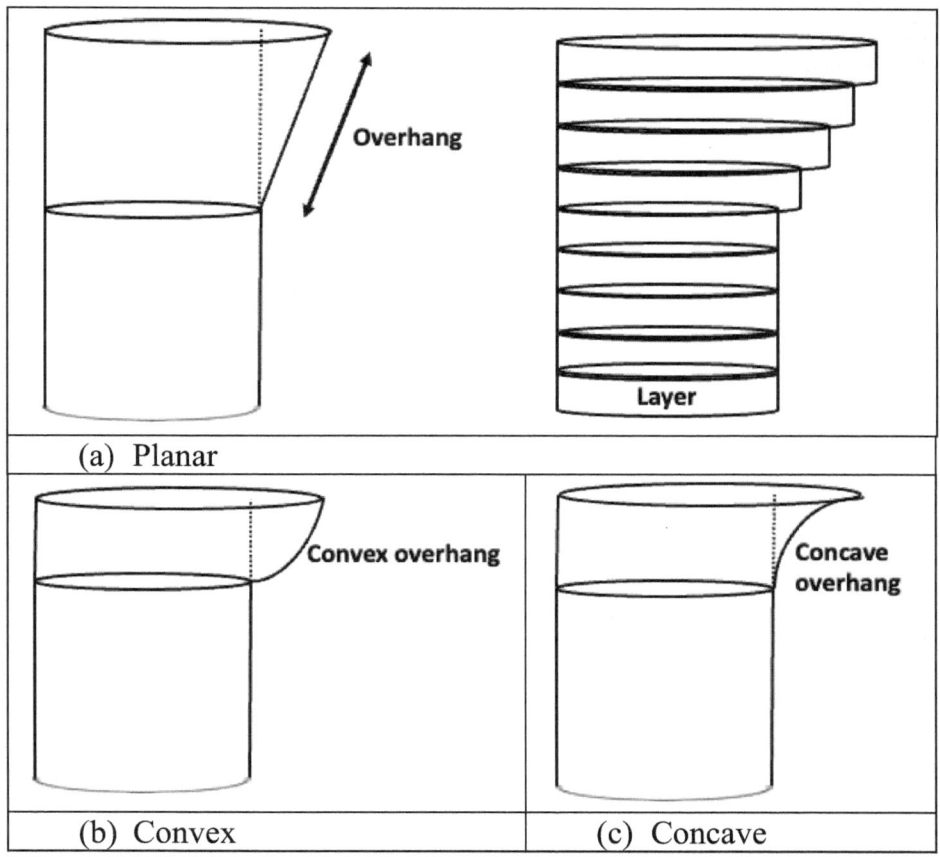

Fig. 2.51   First type of overhang: **a** planar, **b** convex, and **c** concave

perimeter of the last layer, a down-facing surface is made that consists of one layer that is just shifted. The rest of the second cylinder is made on this layer but does not contribute to the formation of this layer. Thus, this type is different from the first type. Because many layers contribute to the formation of the first type.

Figure 2.52b shows the next layer of the first cylinder shifts outside red dotted line to create an overhang and a cylinder. The next layer of the second cylinder again shifts outside the green dotted line to create the second overhang and the third cylinder.

When a layer changes once, one extra cylinder forms, creating one overhang (Fig. 2.52a). If changes occur twice, two extra cylinders form, creating two overhangs of the second type (Fig. 2.52b). Changes occur after an interval of many layers. There-fore, cylinders form. Because a cylinder consists of many layers. What if the same change occurs with each layer? Then the cylinders will not form, but outward shifts will. This

Fig. 2.52 Second type of overhang: **a** two cylinders, and **b** three cylinders

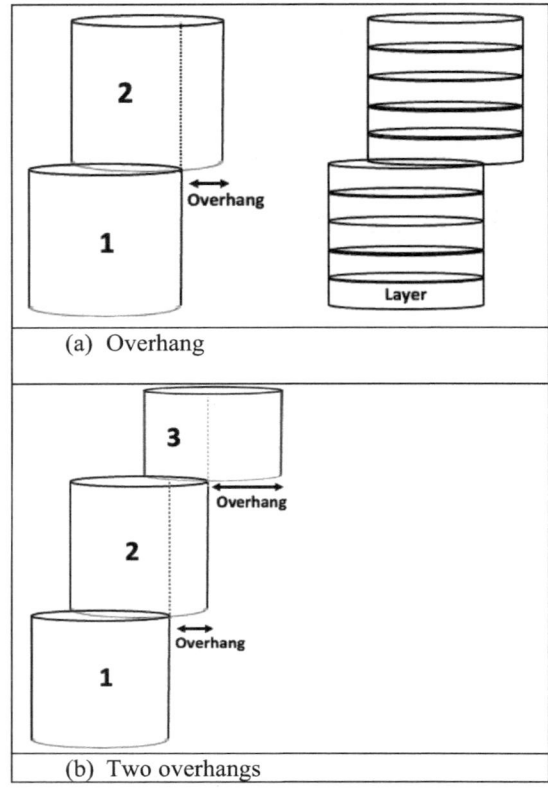

(a) Overhang

(b) Two overhangs

is the same outward shift that is an overhang of the second type. These outward shifts make an angled surface, which is an overhang of the first type. Thus, the conversion of the overhang from the second type to the first type depends upon the interval between two changes. If there is no interval between two changes, the first type forms. Otherwise, the second type forms. The frequency of changes determine which type of surfaces forms. If the overhang of the second type forms, it requires a support structure to manufacture. Manufacturing it may be problem, but this is not an unsolvable problem. If the overhang of the first type forms, the overhang contains the gaps that cannot be filled up, making the problem unsolvable. Thus, there are two types of manufacturing problems (solvable and unsolvable) originating from two types of layer arrangements. One type of layer arrangement is where changes occur without any interval, i.e., changes occur consecutively; another type is where changes occur non-consecutively.

Figure 2.53 shows both types of overhangs. In the first type, the angle may be more than 0° and less than 90° (or more than 90° and less than 180°).

Figure 2.54a shows the design of a circular hole made on the vertical surface of a rectangular block. Which area of the hole is an overhang? The design can be divided in

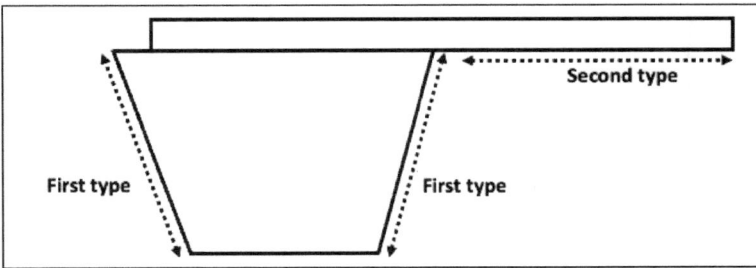

Fig. 2.53 First type and second type of overhang

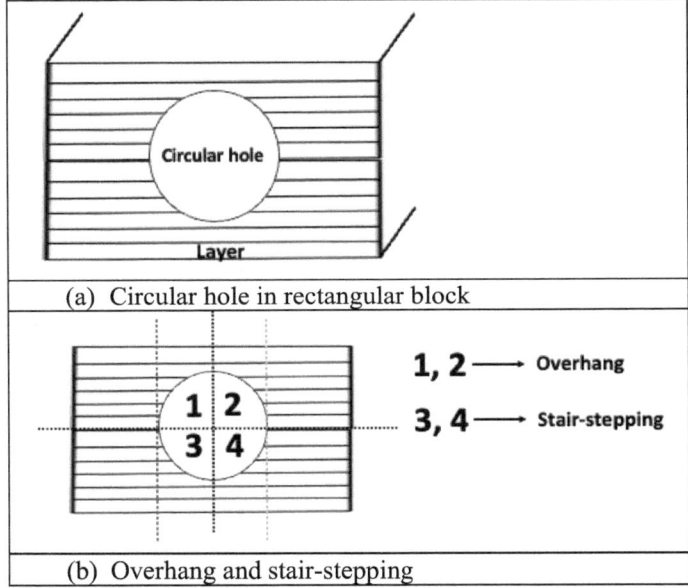

Fig. 2.54 Overhang in circular hole: **a** circular hole, and **b** overhang

four zones. The upper half of the hole consisting of two zones shows overhangs, while the lower half consisting of two zones shows stair-stepping (Fig. 2.54b).

In zone 1, the next layer shifts outside the perimeter of the last layer as seen by the shifting outside the red dotted line, the shifting shows the first type of overhang (Fig. 2.55a). In zone 2, the next layer shifts outside the last layer as seen by the shifting outside the green dotted line, the shifting again shows the first type (Fig. 2.55b). But in zone 3 and 4, the shifting does not take place outside. In zone 3, the next layer shifts inside the perimeter of the last layer as seen by the shifting towards the dotted red line, the shifting shows the stair-stepping (Fig. 2.55c). In zone 4, the next layer shifts inside the

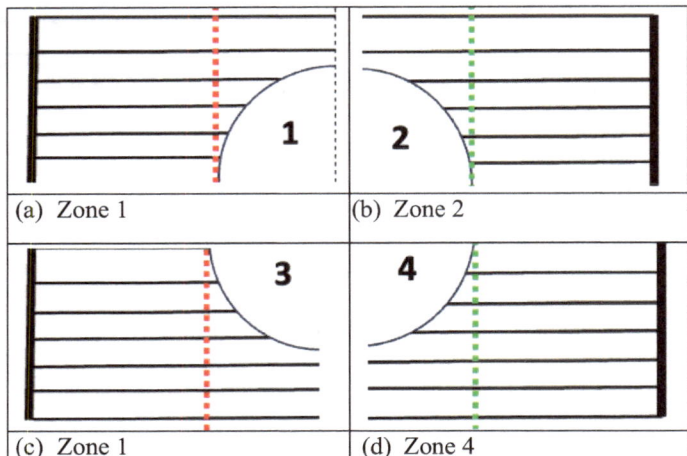

(a) Zone 1   (b) Zone 2

(c) Zone 1   (d) Zone 4

Fig. 2.55 Various zones of circular hole: **a** overhang, **b** overhang, **c** stair-stepping, and **d** stair-stepping

last layer as seen by the shifting towards the green dotted line, the shifting again shows the stair-stepping (Fig. 2.55d).

What if a rectangular hole is made on the vertical surface of a rectangular block? In this hole (Fig. 2.56a), there is no gradual shifting. The absence of the gradual shifting eliminates the possibility of either the first type of overhang or stair-stepping. When the hole is started to be made on a vertical surface, there is no connection between two sides of the hole. At the end of the hole, when the next layer is deposited to end the state of no connection, the next layer does not shift with respect to either the left or the right side of the hole. Because for the shift to take place, there must be a gap in the next layer, i.e., the next layer must be broken. In the absence of the gap, no overhang forms (Fig. 2.56a). While in the case of circular hole, there remains always gaps for the next layer to shift until the layer reaches the end of the circular hole. However, if there is a gap in the next layer within a rectangular hole, overhangs of the second type forms (Fig. 2.56b).

Fig. 2.56 Rectangular hole:
a no overhang, **b** overhangs

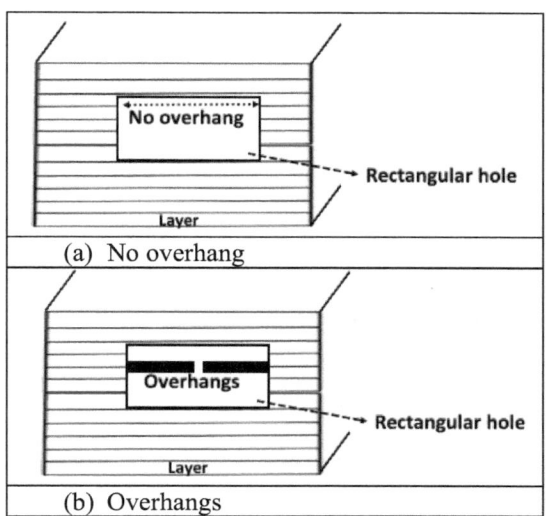

Design Guidelines

3

3.1   Which Hollow Shape Should Be Made on Which Surface?

The advantage of AM is that it allows any shape to be made if the shape is confined only to one layer. For example, if a circle or a rectangle needs to be made, it can be made by drawing it on a horizontal surface.

The problem starts when the shape is not confined to a single layer and the shape starts to shift from an imaginary line that is drawn perpendicular to the layer. For example, for making a cylindrical hole normal to a horizontal plane, one circle is drawn on another circle, and it continues. The position of one circle does not shift with respect to another circle. There is no problem in making the cylindrical hole, because if an imaginary line is drawn on a circle perpendicular to the layer, no circle shifts either towards or away from the line (Fig. 3.1a). If the same cylindrical hole is made on a vertical surface, problems start. If an imaginary line is drawn perpendicular to the layer, the circumference of the circle always shifts from this line. This is a problem, which shows there will be difficulty in making it (Fig. 3.1b).

When a rectangle is made normal to a vertical plane, there is no shift for two of its sides that are perpendicular to the layer, which shows there will be no problem in making its two sides. For making its lower side, there is no problem because this line is confined to a single layer, i.e., the first layer. Similarly, for making its upper side, there is no problem because this line is confined to a single layer, i.e. the last layer. A rectangle made using twenty layers is shown (Fig. 3.2).

To avoid the problem due to shift, if a circle needs to made, it must not be made on a vertical plane. It should be made on a horizontal surface. If a rectangle needs to be made, it can be made either on a vertical or a horizontal surface. In a vertical surface, it can be made because there is no shift from the normal line. In a horizontal surface, it can be made because it is confined to a single layer.

© The Author(s), under exclusive license to Springer Nature Switzerland AG 2025   61
S. Kumar, *A New Theory of Additive Manufacturing*, Synthesis Lectures on Engineering, Science, and Technology, https://doi.org/10.1007/978-3-031-75427-2_3

Fig. 3.1 Shifts in cylindrical
hole

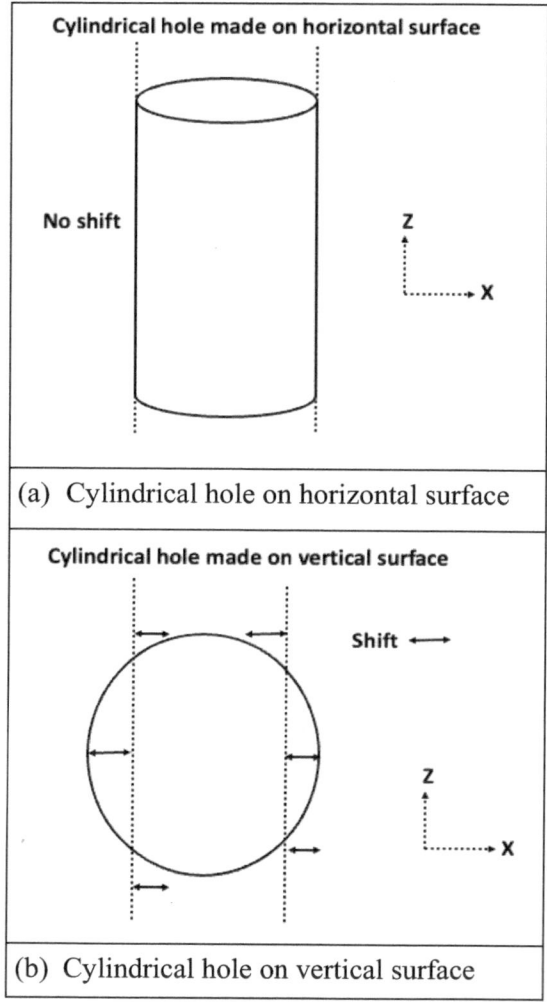

If a cylindrical hole needs to be made, it can be made on a horizontal surface, i.e., normal to a horizontal plane, because there will be no shift from a normal line. If it is made on a vertical surface, i.e., normal to a vertical plane, there will be a shift, which will give a problem. A rectangular hole can be made on either surface because there will be no shift in either case (Fig. 3.3).

What if the diameter of a cylindrical hole changes? It will give problems on either surface because there will be shifts. Thus, there will be problems in making a conical hole normal to either surface (Fig. 3.4). What if a rectangular hole becomes increasingly narrow? It will give problems on either surface because there will be shifts. Thus, there will be problems in making a trapezium hole normal to either surface (Fig. 3.5).

Fig. 3.2 Rectangle made
using twenty layers

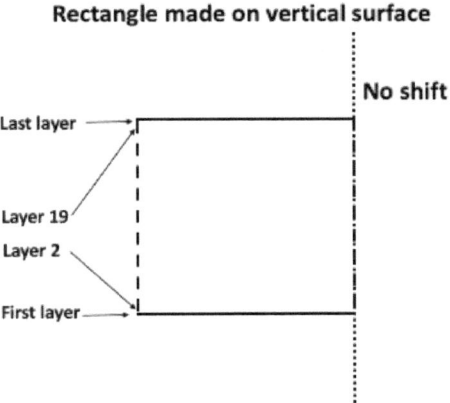

What the theory says about making various hollow shapes? It says a shift from a normal imaginary line is a problem. The shapes that show the shift should not be preferred. For example, a cylindrical hole shows shifts when made normal to a vertical plane, while it does not show when made normal to a horizontal plane; therefore, a cylindrical or circular hole should be made on a horizontal plane (Fig. 3.6).

The theory does not find a difference between making a rectangular or square hole on either surface. Therefore, there is no preference between a horizontal and vertical surface for making it (Fig. 3.7). However, in practice, the vertical surface is not preferred because it requires support structure to make the hole normal to the vertical surface. The theory says the accuracy of a hole does not depend upon the support structure. If it depends, it is because there is a problem in making the support structure, which causes the accuracy to depend upon how well the support structure is made. Therefore, if a design requires this hole to be made normal only to a vertical surface, attention should be given on how the support structure should be made and not on how to avoid making it by changing the design.

The theory does not find a difference between making a rectangular or square hole and a cylindrical or circular hole on a horizontal surface (Fig. 3.8). However, in practice, a circular hole is preferred over a square hole. The reason for this preference is that a square hole requires a sharp corner to be made, while the circular hole is free from that requirement. Since making a sharp corner is difficult in machining, it is assumed that this will also be difficult in AM. The theory says making both holes are either equally difficult or equally easy, and therefore, one should not be preferred over another. If an AM technique finds it easier to make one than another, it could be due to shortcomings of the technique that it cannot make both with equal ease.

Fig. 3.3 Shifts in rectangular hole

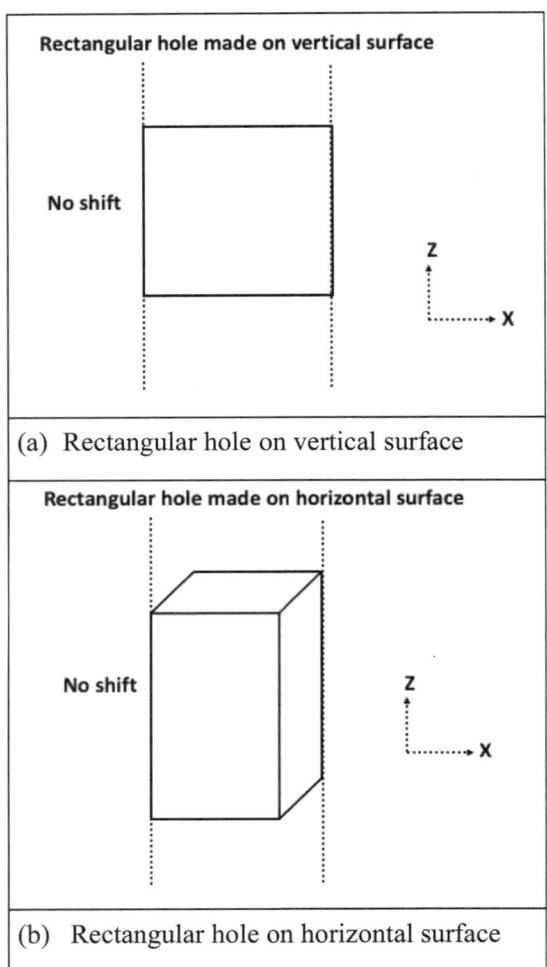

Rectangular hole made on vertical surface

No shift

(a)   Rectangular hole on vertical surface

Rectangular hole made on horizontal surface

No shift

(b)   Rectangular hole on horizontal surface

3.2   Upright Cylinder Is Better than Horizontal Cylinder

Which orientation of a cylinder is better to make: upright (Fig. 3.9a) or horizontal (Fig. 3.9b)? As per the theory, the upright orientation, should be preferred over the horizontal orientation, as the former will give a more accurate part.

While in practice, the horizontal orientation is preferred. It needs to be rechecked why this practice is. This orientation is preferred because the height is smaller in this orientation, meaning a decreased number of layers are required to fabricate, which accelerates fabrication. The theory says those designs that accurately match the layer arrangement will furnish accurate fabrication. If they do not match, optimization of the fabrication is

Fig. 3.4 Shifts in conical hole

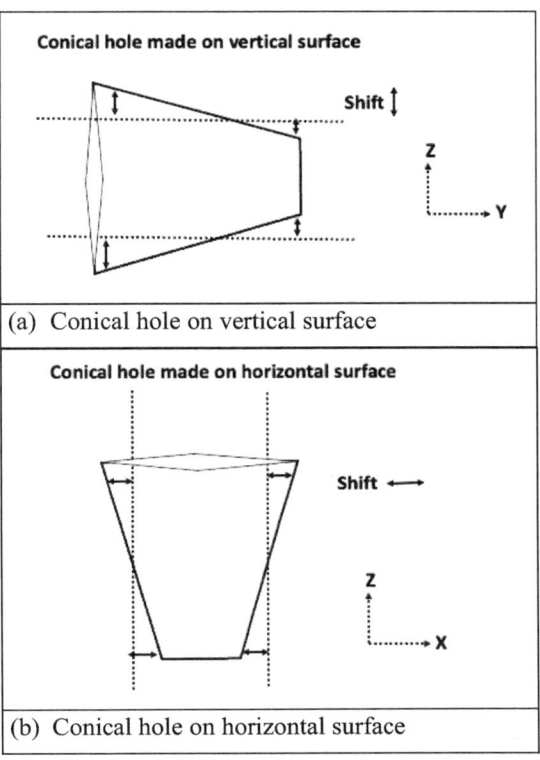

(a) Conical hole on vertical surface

(b) Conical hole on horizontal surface

required to reach near the accurate fabrication, causing delay. Therefore, the horizontal orientation, where the layer arrangement does not accurately match the design, will need more time to find optimized values to fabricate. The time needed will offset the perceived gain in time that comes from a decreased number of layers. Therefore, if this orientation is used, it will not accelerate but delay fabrication.

If it is said it accelerates fabrication and the result of the fabrication is not unacceptable, this is based on a misconception. The misconception is when a cylinder is horizontal, it requires a decreased number of layers because its height is smaller. The number of layers is not decided by the height but by the degree of curvature. If the curve of a design bends more, it requires an increased number of layers to decrease the effect of the bent on accuracy. If, in practice, the number of layers for a horizontal orientation does not increase, it is because there is no change in the layer thickness for both orientations. Because the layer thickness is not based on the degree of curvature but on the ability of a layer to be converted from raw material to consolidated material. The lack of conversion exposes the limitation of AM. Due to the limitation, the layer thickness cannot be decreased for a horizontal orientation—a very thin layer cannot be made. Again, due to the limitation, the

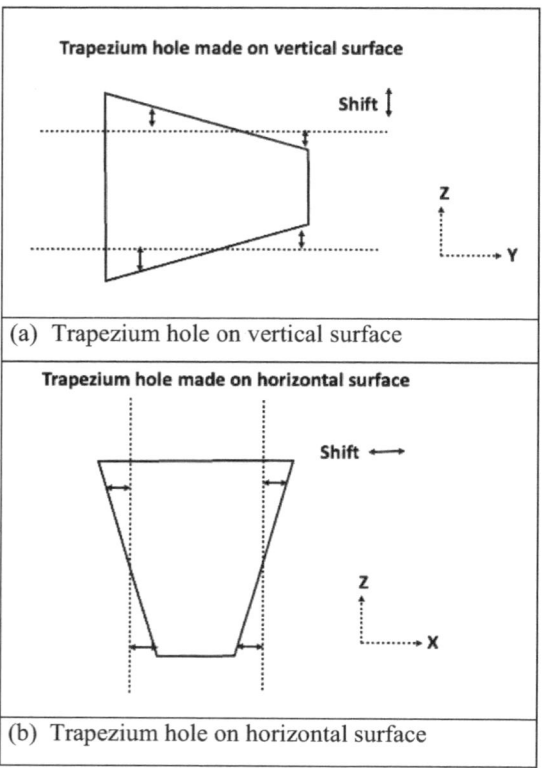

Fig. 3.5   Shifts in trapezium hole

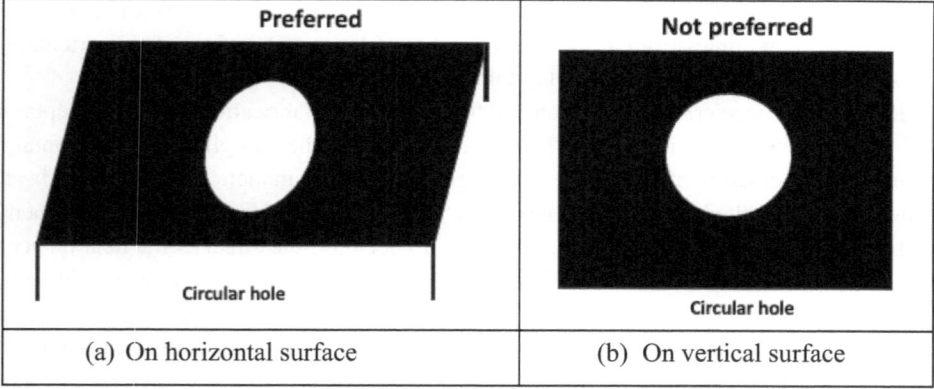

Fig. 3.6   Circular hole on horizontal and vertical surface

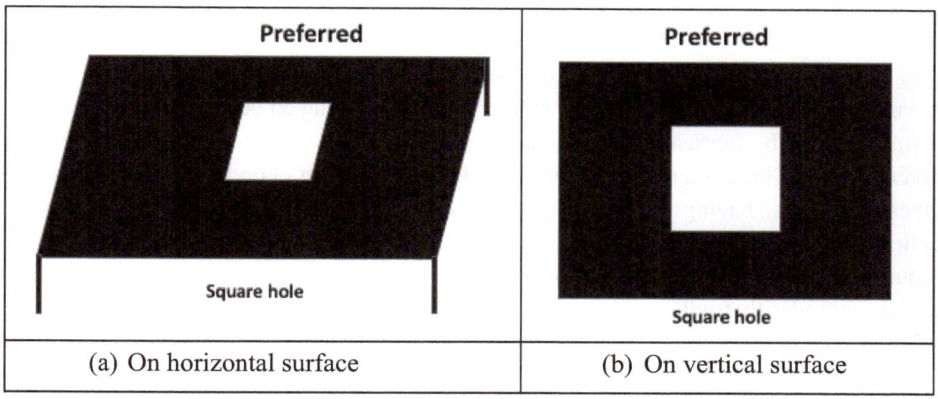

Fig. 3.7   Square hole on either surface

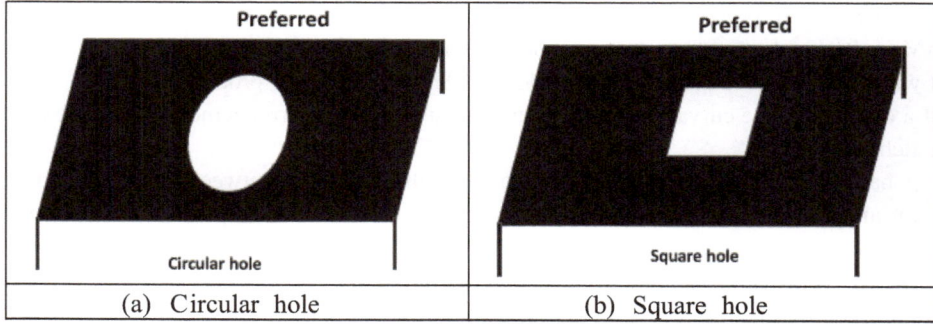

Fig. 3.8   Square and circular hole on horizontal surface

Fig. 3.9   Upright and horizontal cylinder

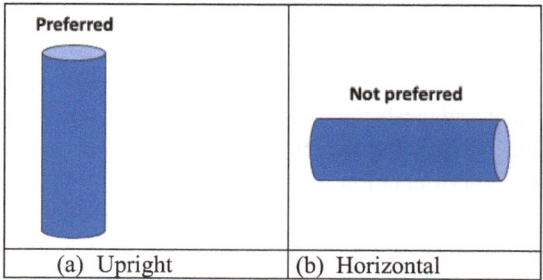

layer thickness cannot be increased for an upright orientation—a very thick layer cannot be made.

This limitation does not allow to check where the accelerated fabrication lies. When it is said the fabrication of a horizontal cylinder should be delayed because it has a curvature, it implies that the fabrication of an upright cylinder should be accelerated because it has no curvature. What if there is no limitation? Then the upright cylinder will require just one layer, i.e. a layer having thickness equal to the height of the cylinder. And the horizontal cylinder will require many layers. Therefore, if a cylinder is selected depending upon who requires fewer layers, it will be the upright one and not the horizontal one. Thus, it is the upright cylinder that will accelerate the fabrication.

If a horizontal cylinder is preferred, this is a practice due to limitations, i.e., the inability to make thinner or thicker layers. These limitations do not come from the principle of AM. Moreover, these limitations do not allow AM to be implemented well. They are the limitations of material processing, machine architecture, machine control, etc. They are not less important because they come from somewhere else. It only means if they are considered main problems of AM, it will not allow to know what the primary problem of AM is. It also does not mean they should not be solved. Because even if they are solved, it will not help solve the primary problem of AM. The primary problem is the fabrication of a curvature. The curvature is a problem because the layer arrangement does not exactly match it.

Changing the orientation is the way to avoid the mismatch between the layer arrangement and the curvature. When the orientation changes from Fig. 3.9b to a, the problem is solved. It is solved by choosing a layer arrangement that does not create gaps. It is not solved by filling up the gaps, because it is not possible. The problem of filling the gaps is managed by making layers thin, so the gaps will be smaller.

When the theory suggests an upright orientation should be preferred, it is because it is the only way how a precision part can be made. The limitations of AM have minimal effect in this orientation. If AM has a limitation that it cannot make a thin layers, this orientation does not require a thin layer to achieve accuracy. This orientation does not demand that AM first be free from its limitation for it to furnish an accurate part. The theory suggests that if the horizontal orientation provides fast fabrication, this merit should not conceal the fact that this orientation is not capable to give as accurate part as the upright orientation gives.

When an upright orientation does not demand a thinner layer to achieve accuracy, it does not question the usefulness of thinner layers. It says the purpose for which a thinner layer is required in a horizontal orientation, it is free from that purpose. If an accurate cylinder is required, the upright orientation is not as helpless as the horizontal orientation when AM cannot make thinner layers.

If a rectangular block instead of a cylinder needs to be made, it can be made in either orientation (Fig. 3.10). If a complex design consists of rectangular blocks, the design is free to select a number of orientations of the design because both orientations of the

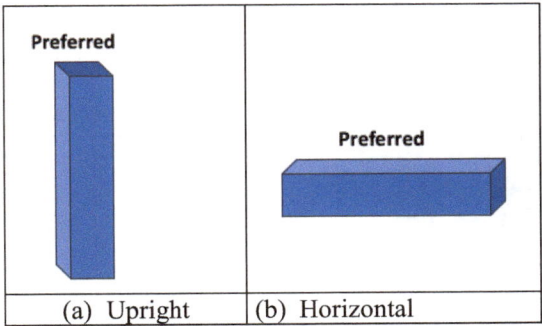

Fig. 3.10   Preferred orientations of rectangular block

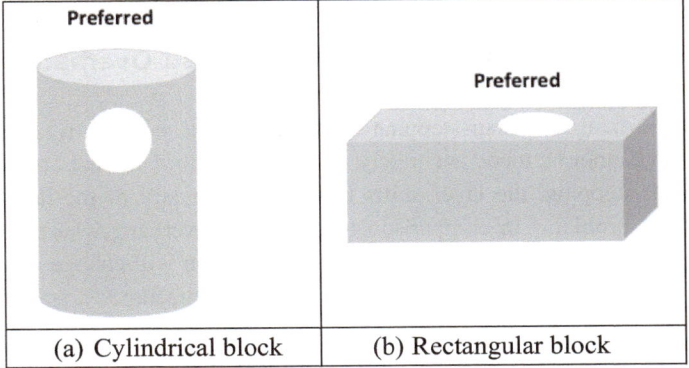

Fig. 3.11   Preferred orientations of blocks having circular holes

blocks are preferred. This freedom will allow to change the orientation of the design depending on which orientations of other features of the design are convenient to make.

Which orientation of a block is preferred when the block has a circular hole? The circular hole needs to be made on a horizontal surface, while a cylindrical block needs to be made in an upright orientation. Therefore, if any of horizontal and upright orientations is chosen, either the block or the hole cannot be made well. Since the block is bigger, the orientation of the block needs to be preferred over the accuracy of the hole. Therefore, the upright orientation should be preferred (Fig. 3.11a). For a rectangular block, either orientation works well, therefore the horizontal orientation of the block, which is also the preferred orientation of the hole, needs to be preferred (Fig. 3.11b).

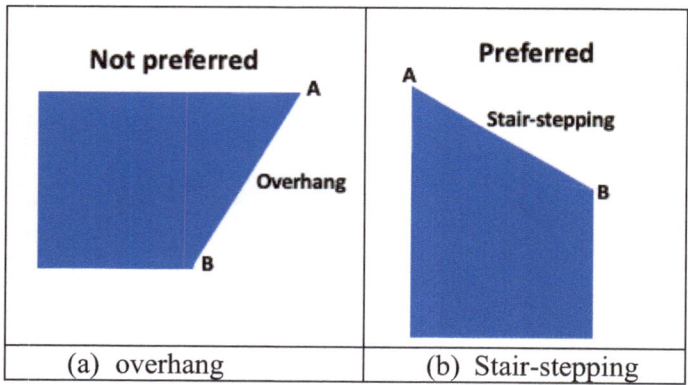

Fig. 3.12 Overhang and stair-stepping

3.3 Stair-Stepping Is Preferred over Angled Overhang

Both angled overhang and stair-stepping are made by the consecutive shifts of layers. Therefore, both cannot be made accurately, which does not make them preferred equally. Because in stair-stepping, the layer shifts inside the boundary of the last layers, which will require less problems in comparison to an angled overhang where the layer shifts outside. Thus, if geometry allows, changing the orientation will change an overhang to a stair-stepping (Fig. 3.12). An overhang (AB, Fig. 3.12a) becomes a stair-stepping (AB, Fig. 3.12b).

3.4 Horizontal Overhang Is Better than Angled Overhang

Horizontal overhang may or may not need a support structure, but if everything goes well, it can be made perfectly, while the angled overhang cannot be made (Fig. 3.13). Therefore, a horizontal overhang is preferred over an angled overhang. This is the same reason why a horizontal bridge should be preferred over an oblique bridge (Fig. 3.14).

3.5 No Preference Between Horizontal Overhang and Horizontal Bridge

The requirement of support structure for making horizontal overhang or bridge depends upon their size and the type of AM process (Fig. 3.15). If their sizes are small, the support structure will not be required in bed process unless a low surface roughness of the underlying surface is required. In deposition process, for a critical size of the bridge,

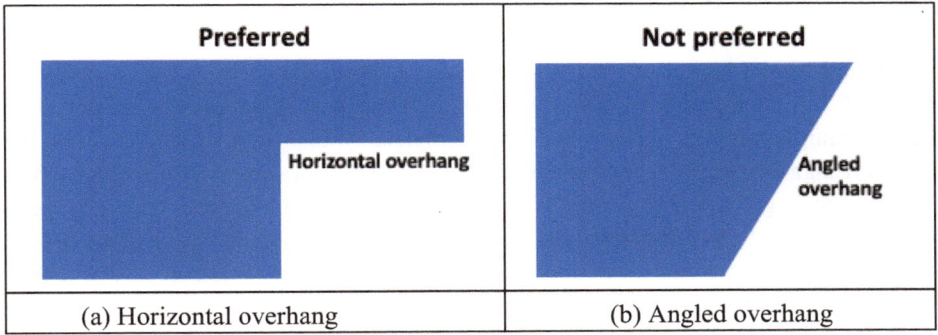

Fig. 3.13 Horizontal and angled overhang

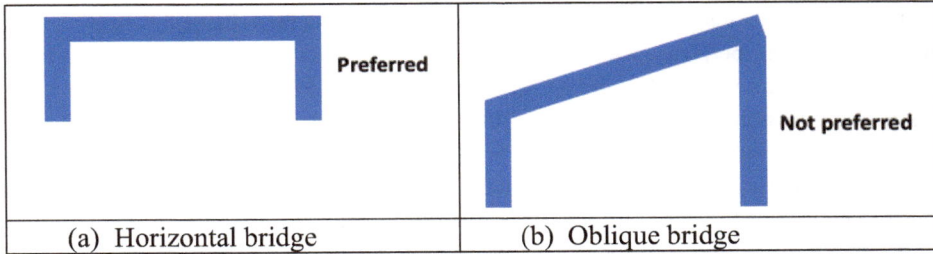

Fig. 3.14 Horizontal and oblique bridge

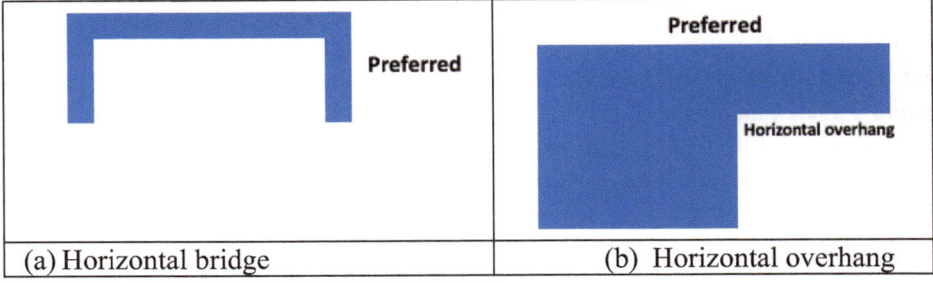

Fig. 3.15 Horizontal bridge and overhang

the pillar of the bridge will act as a support—this facility is not available in the case of an overhang. Therefore, for a critical size, the bridge can be made better than the overhang. However, if their sizes are big, the support structure will be required in all AM techniques.

3.6   How to Select an Orientation?

If a design cannot be changed, it can be oriented. What the theory says about the orientation? The theory says a shift is a problem. Therefore, if an orientation that avoids the maximum number of shifts is the best. Complete avoidance of shifts is possible only when a design is made by the extension of a sketch that is drawn on a horizontal surface.

A design is given in Fig. 3.16. Its shape is covered by six surfaces: two rectangular planes (ADEC, ADFB), two triangular planes (ABC, DFE), and one square plane (BCEF). The design creates manufacturing problems in a number of orientations (Fig. 3.17). When the design is oriented so its square plane is on the horizontal surface, there are stair-stepping problems (Fig. 3.17a). When this orientation is reversed by 180° (Fig. 3.17b), there are overhang problems. When the rectangular plane of the design is on the horizontal surface, there are stair-stepping problems (Fig. 3.17c, e). When the orientations are reversed by 180° (Fig. 3.17d, f), there are overhang problems.

The theory says manufacturing problems happen because there is a shift from an imaginary line passing perpendicular to a horizontal surface. For example, there is a problem in Fig. 3.17a because line AB and AC are bent. There would not have been any problems if AB and AC were parallel to z-axis. But if they were parallel, the design would have been different. But it shows the way how to avoid problems. Is there any possibility to orient the design so its extension will be parallel to z-axis? This can be tried. For this, various projections of the design needs to be taken, and these projections need to be redrawn on a horizontal surface, and then these projections need to be extended along z-axis. If after the extension, the extended geometry matches the given design, this will be the right orientation.

Figure 3.18 shows six projections of the design on six planes: top, bottom, left, right, back, and front. Out of these six projections, three projections are symmetrical and they are redundant. Therefore three projections: top view (Fig. 3.18a), left view (Fig. 3.18c), and back view (Fig. 3.18e) need to be tried by drawing them on horizontal surfaces. Figure 3.19a, c, and e shows these projections are respectively drawn on horizontal surfaces. Figure 3.19b, d, and f shows their respective extensions. If a rectangle will be extended, it will become a rectangular block (Fig. 3.19d). If a square will be extended, it

Fig. 3.16 A design to be oriented

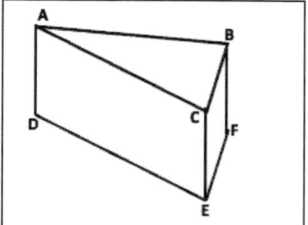

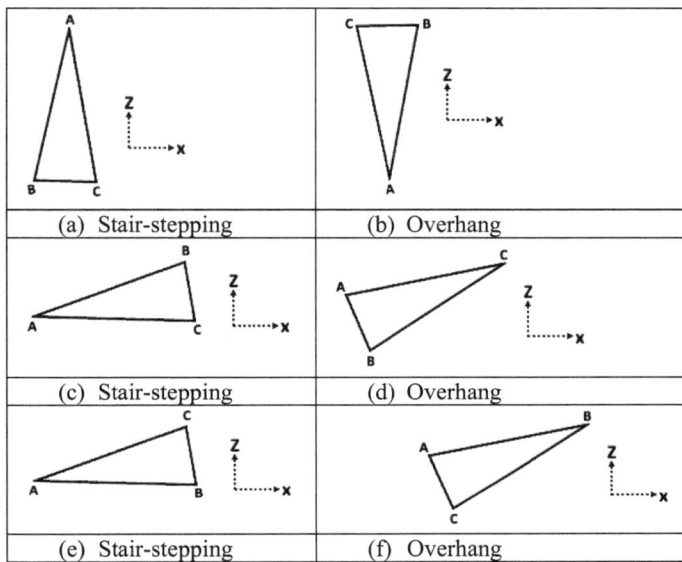

Fig. 3.17 Problems due to orientations: **a** stair-stepping, **b** overhang, **c** stair-stepping, **d** overhang, **e** stair-stepping, and **f** overhang

will become a square (Fig. 3.19f) or rectangular block. If a triangle will be extended, it will become a triangular block (Fig. 3.19b) that is the design.

Therefore, if the design (Fig. 3.16) is oriented to make its top face parallel to the horizontal surface, it will lead to manufacture a part without having stair-stepping and overhang problems. The example of design (Fig. 3.16) is intentionally chosen so at least one of its projections will help make part without problems, which will show how to select an orientation. But it is difficult to find such unique orientation for all complex designs. However, the theory will help find a preferred orientation, out of many orientations for a complex design, that will create less manufacturing problems.

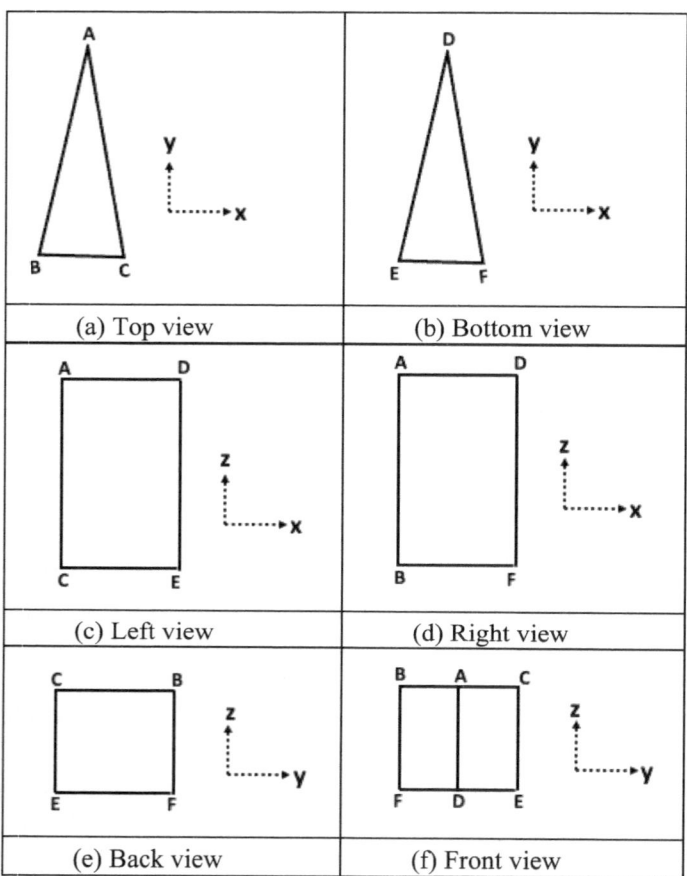

Fig. 3.18 Projections of design on various planes

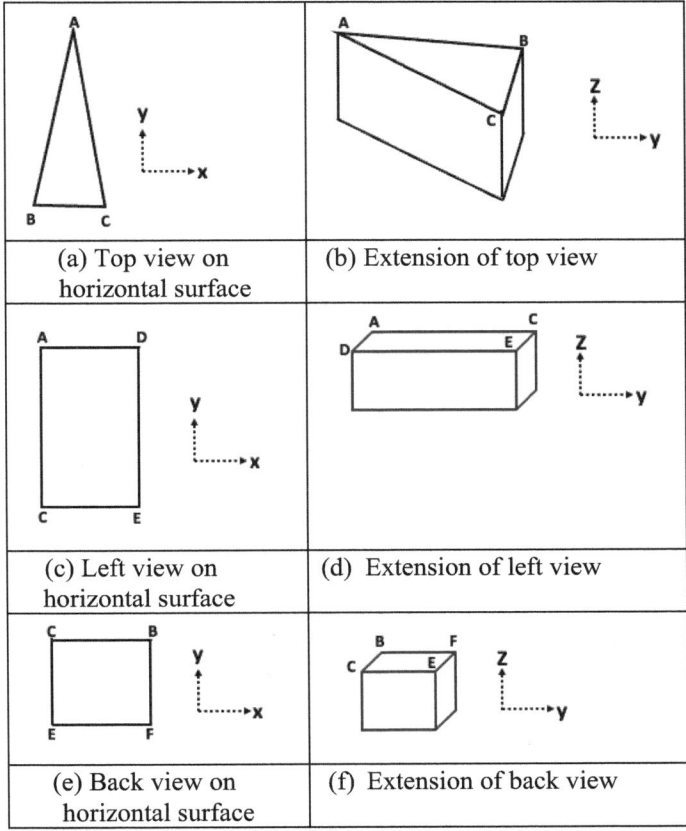

(a) Top view on horizontal surface	(b) Extension of top view
(c) Left view on horizontal surface	(d) Extension of left view
(e) Back view on horizontal surface	(f) Extension of back view

Fig. 3.19 Projections drawn on horizontal surface and their extensions